Otherness, Othering and Space in Tourism Futures

THE FUTURE OF TOURISM

Series Editors: **Ian Yeoman**, *NHL Stenden University of Applied Sciences, the Netherlands* and **Una McMahon-Beattie**, *Ulster University, Northern Ireland, UK*

Some would say that the only certainties are birth and death; everything else that happens in between is uncertain. Uncertainty stems from risk, a lack of understanding or a lack of familiarity. Whether it is political instability, autonomous transport, hypersonic travel or peak oil, the future of tourism is full of uncertainty but it can be explained or imagined through trend analysis, economic forecasting or scenario planning.

The Future of Tourism sets out to address the challenges and unexplained futures of tourism, events and hospitality. By addressing the big questions of change, examining new theories and frameworks or critical issues pertaining to research or industry, the series will stretch your understanding and generate dialogue about the future. By adopting a multidisciplinary perspective, be it through science fiction or computer-generated equilibrium modelling of tourism economies, the series will explain and structure the future – to help researchers, managers and students understand how futures could occur. The series welcomes proposals on emerging trends and critical issues across the tourism industry and research. All proposals must emphasise the future and be embedded in research.

All books in this series are externally peer-reviewed.

Full details of all the books in this series and of all our other publications can be found on http://www.channelviewpublications.com, or by writing to Channel View Publications, BLOCK, The Fairfax, Pithay Court, Bristol, BS1 3BN, UK.

THE FUTURE OF TOURISM: 13

Otherness, Othering and Space in Tourism Futures

Edited by
Lucia Tomassini

CHANNEL VIEW PUBLICATIONS
Bristol • Jackson

To Rudi, my non-human soulmate — with love and gratitude, forever.

DOI https://doi.org/10.21832/TOMASS9325
Library of Congress Cataloging in Publication Data
A catalog record for this book is available from the Library of Congress.
Names: Tomassini, Lucia, editor
Title: Otherness, Othering and Space in Tourism Futures/Edited by Lucia Tomassini.
Description: Bristol; Jackson: Channel View Publications, [2026] | Series: The Future of Tourism: 13 | Collection of essays by Elisa Burrai and 23 others. | Includes bibliographical references and index. | Summary: "This book examines the concepts of otherness and othering in tourism, investigating how human and non-human actors are perceived within tourism spaces. It challenges traditional views, emphasising relationality and justice, and integrates future studies and theories to explore emerging trends, critical reflections and future scenarios for tourism"—Provided by publisher.
Identifiers: LCCN 2025021406 (print) | LCCN 2025021407 (ebook) | ISBN 9781845419325 hbk | ISBN 9781845419332 pdf | ISBN 9781845419349 epub
Subjects: LCSH: Culture and tourism | Tourism—Social aspects | Sustainable tourism | Animal welfare
Classification: LCC G156.5.H47 O84 2026 (print) | LCC G156.5.H47 (ebook)
LC record available at https://lccn.loc.gov/2025021406
LC ebook record available at https://lccn.loc.gov/2025021407

British Library Cataloguing in Publication Data
A catalogue entry for this book is available from the British Library.

ISBN-13: 978-1-84541-932-5 (hbk)

Channel View Publications
UK: BLOCK, The Fairfax, Pithay Court, Bristol, BS1 3BN, UK.
USA: Ingram, Jackson, TN, USA.
Authorised Representative: Easy Access System Europe – Mustamäe tee 50, 10621 Tallinn, Estonia, gpsr.requests@easproject.com.

Website: https://www.channelviewpublications.com
X: Channel_View
Bluesky: @channel-view.bsky.social
Facebook: https://www.facebook.com/channelviewpublications
Blog: https://www.channelviewpublications.wordpress.com

The policy of Multilingual Matters/Channel View Publications is to use papers that are natural, renewable and recyclable products, made from wood grown in sustainable forests. In the manufacturing process of our books, and to further support our policy, preference is given to printers that have FSC and PEFC Chain of Custody certification. The FSC and/or PEFC logos will appear on those books where full certification has been granted to the printer concerned.

Typeset by SAN Publishing Services.

Contents

Contributors

Mikko Äijälä is a Postdoctoral Researcher at the Faculty of Social Sciences, University of Lapland, Finland. He is curious about developing a multispecies approach to the study of landscapes. Currently, he is working on the DEATNU project to find out the local, indigenous ways of knowing salmon, managing salmon fishing, and identifying how this can be integrated into salmon policy and administration for the Teno River, Finland. Previously, he has carried out research into human-sled dog-place relations in the tourism context at the University of Lapland, where he holds a PhD in tourism research. His paths intersect with non-humans in trying to get some answers to these problems, and also beyond academia, living an outdoor life with his dogs.

Jessica Bell Rizzolo is a Postdoctoral Researcher in the Department of Fisheries, Wildlife and Conservation Sciences at Oregon State University. She holds a joint PhD in Sociology and Environmental Science and Policy from Michigan State University. Her research interests include wildlife tourism, wildlife consumption, and the legal and illegal wildlife trades, particularly as related to elephants and big cats. Her work has appeared in journals such as *Global Ecology and Conservation*, *Nature Conservation*, *EcoHealth* and the *Journal of Sustainable Tourism*.

Giovanna Bertella is Professor at the School of Business and Economics, UiT – The Arctic University of Norway, Tromsø. She received her PhD about learning and networking in tourism from the Department of Sociology, Political Science and Community Planning at UiT. Her research interests are management, entrepreneurship/innovation, marketing, tourism, and leisure studies (nature- and animal-based experiences, rural tourism, food tourism, events), food studies (plant-based).

Elisa Burrai (PhD) is Senior Lecturer in the School of Events, Tourism and Hospitality Management at Leeds Beckett University (UK). Elisa's research focuses on the intersection between tourism, responsibility, and ethics. Her research interests include interdisciplinary approaches to volunteer tourism, global development, tourism ethnographies and development in the Global South. More recently, Elisa's research interests have focused on the nexus between tourism and (forced) migration.

Guillaume Ethier is Professor in Urban Studies at the *École des Sciences de la Gestion* of the Université du Québec à Montréal, Canada (UQAM). His research focuses on placemaking and contemporary urban theory. In 2023, he published his latest essay on cities in the urban age, *La Ville Analogique* (Éditions L'apogée).

David A. Fennell's main focus of research is on tourism ethics and ecotourism. He has published widely in these areas, including books on *Ecotourism* (5th edn), *Tourism Ethics* (2nd edn), *Codes of Ethics in Tourism, Tourism and Animal Ethics* (2nd edn) and *Sustainable Tourism: Principles, Contexts and Practices.* A major thrust of his research involves theory from other disciplines to gain traction on tourism's most persistent problems. Fennell is the founding Editor-in-Chief of the *Journal of Ecotourism*, is the editor of a Routledge book series on tourism ethics and is the editor of two Routledge Handbooks: *Tourism and the Environment* and *Ecotourism.*

Sarah Gainsforth (1980) is an independent researcher, a writer and freelance journalist; her work focuses on housing.

José-Carlos García-Rosell is Professor of Business Ethics at Oulu Business School, University of Oulu, Finland. Additionally, he holds the position of visiting scholar within the Faculty of Social Sciences at the University of Lapland's Multidimensional Tourism Institute (MTI). He also serves as an associate professor in corporate social responsibility in tourism at Tampere University, Finland and the University of Maribor, Slovenia. He works in the fields of corporate social responsibility, business ethics, responsible tourism business and responsible management education. His current research agenda focuses on human-animal relations, responsible tourism experiences and experiential learning.

Chiara Grasso ethologist specialized in human-animal relationships in tourism activities. Since 2018 he has been president of ETICOSCIENZA Association. She obtained certification as an environmental guide in Italy and a safari guide in South Africa (FGASA). She is currently involved in scientific journalism and environmental education.

Minni Haanpää (PhD) works as a Senior Lecturer in tourism studies at the Multidimensional Tourism Institute (MTI), Faculty of Social Sciences, University of Lapland, Finland. Her research interests include tourism and event experiences, affect, value co-creation, and ethnographic methodologies, especially videography and autoethnography. Her current research projects explore the possibilities of tourism as social change in the Nordic urban context and technology-mediated tourism experiences.

Hindertje Hoarau-Heemstra is Associate Professor in Management and Organisation at the faculty of Social Sciences at Nord University, Bodø, Norway. She holds a PhD in business economics from Nord University Business School. She explores the links between organisations, local communities and other-than-human animals from a posthuman perspective. Her main research topics are related to innovation, management, sustainability and the development of local communities in the context of tourism.

Dominic Lapointe is Professor in the Department of Urban and Tourism Studies at Université du Québec à Montréal (UQAM), Canada. He holds the *Chaire de recherche sur les dynamiques touristiques et les relations socioterritoriales* and leaders of the *Groupe de recherche et d'intervention tourisme territoire et société* (GRITTS) at UQAM. His work explores the production of tourism space and its role in the capitalist system expansion and its biopolitical dimensions. Its latest research looks at climate change, social innovations, indigeneity and critical perspective in tourism studies.

Laurence Laroque Labossière is a master's student at the Department of Geography of the Université du Québec à Montréal, Canada (UQAM). With a bachelor's degree in International Studies and Modern Languages from the Université Laval du Québec, Canada, her research interests focus on tourism and the socio-territorial transformations that come with it.

Christian Lenzi is the co-founder and vice-president of the association ETICOSCIENZA, Italy. Wildlife biologist specialized in animal behavior. He has conducted environmental volunteering in Italy and abroad and was a wildlife manager in a Costa Rican rescue centre. He is currently involved in nutrition (human and animal) and the management of a small ethical farm.

Ludovica Lombardini is co-founder and co-owner of *La Tana del Bianconiglio*, an agrotourism located close to the Italian town of Siena, in Tuscany. *La Tana del Bianconiglio* is also a shelter for free animals that is part of the Sanctuary Network, a non-profit organization born with the aim of welcoming and rescuing animals of all species mainly considered 'for profit', whose short existence is destined to end inside a slaughterhouse. Ludovica works with all aspects of the agritourism: agricultural production, tourism, and the care of the animals hosted at the shelter.

Michal Lovecky is a founder and creative director specializing in immersive media production at the nexus of technology, arts and science. Holding Master's degrees in Business Administration and Arts, he is currently pursuing a third Master's in the interdisciplinary program Sea and Society at the University of Gothenburg. His research focuses on leveraging digital technologies to foster collective action and stewardship for sustainable ocean futures. He is currently developing a digital underwater art gallery and leads his companies – *Seanergy, Cyan Planet* and *Go360* – utilising immersive media for purposes ranging from corporate training to enhancing nature connection. His projects extensively explore humanity's relationship with the ocean.

Julien Menu is a PhD student in Urban/Tourism studies at Université du Québec à Montréal, Canada (UQAM). His research interests focus on rural areas, pop culture, content tourism and Japan. His thesis questions the rurality of contents tourism as a site of interaction between rural and urban everyday practices in the Japanese context.

Jana Milovanovič works as the Director of *Terra Vera* NGO. She is an anthropologist, researcher, mentor, and consultant. Since 2018, she has been actively involved with the European Migrantour network as the coordinator of Migrantour Ljubljana and Rural Migrantour. Situated as an anthropologist at the complex border intersection between Slovenia and Croatia, she is driven to document and respond to the circumstances affecting the lives of border communities and those crossing them. Her primary research interests include borderlands, migration, ethnography of routes, walking practices, counter-mapping, resourcefulness and peripheral heritage. She brings extensive expertise in international, intercultural, and inclusive projects.

Ninja Mueller is pursuing her PhD in Marine Education Science and investigates decolonising the marine curriculum and applying a critical ocean pedagogy at the intersection of ocean literacy and critical animal studies. Ninja Mueller holds a bachelor's degree in Environmental Sciences, a master's degree in Marine Biological Resources, and is currently finishing a second master's degree in Education for Sustainable Development. She is the co-founder of the immersive media studio *Cyan Planet,* creating immersive experiences for ocean protection. Furthermore, she is an active member of the *Asian Marine Educators Association* (AMEA) Youth Board and member of the *Critical Animal Studies* group of Gothenburg University (GU-CAS).

Meghan L. Muldoon is Assistant Professor in Sustainable Tourism at the University of Groningen (RUG), Campus Fryslan, Leeuwarden, Netherlands. Her research interests and expertise focus on the areas of gender and feminisms, decolonization, and poverty, through the lens of tourism. Also interested in qualitative approaches to research, particularly in arts-based and decolonizing methodologies.

Arianne Simard-Picard graduated in *Gestion et pratiques socioculturelles de la gastronomie* and is currently pursuing a master's degree in *Développement du Tourisme* at the Université du Québec à Montréal, Canada (UQAM). Her research focuses on food studies, examining its role in expressing national identity through sociocultural changes within traditional cuisine.

Lucia Tomassini (PhD) is Senior Research Lecturer at NHL Stenden University of Applied Sciences, Leeuwarden, Netherlands. Her research interests and expertise embrace circular economy and circularity in the physical and sociological space of tourism, posthuman approaches in more-than-human societies, and he interplay between mobilities, spatial theories, and cultural geographies in the context of tourism and leisure.

Francesco Vietti is an anthropologist at the University of Turin. He has carried out extensive fieldwork in Eastern Europe and the Mediterranean, where he studied the intersections of migration, tourism, and cultural heritage. On these topics he published the book *Unexpected Encounters: Tourists and Migrants in the Mediterranean* (Berghahn, 2024). He is interested in the possible dialogue between anthropology and science fiction as a tool for social innovation. In 2023, together with Fabio Dei and Fabiana Dimpflmeier he edited the special issue *Anthropology and Science Fiction* of the *Rivista di Antropologia Contemporanea* (in English, Journal of Contemporary Anthropology).

Anne Wally Ryan is Associate Professor in Geography at Nord University, Bodø, Norway. She holds a PhD from NTNU, Norwegian University of Science and Technology. Her main research topics are related to the phenomenological perspective on experiences, experience-based value creation, contextual issues and place development, and sustainable destination management.

The Future (Copyright: The Academic Line)

Foreword

The other is someone or something who is not me.

But we can't make it alone in the world, so I need to choose some allies, friends even, who I can trust to provide for me the social, emotional, pragmatic, intellectual and economic needs I have. And I'll give back to them, too, because social exchanges work best that way in order to get the most out of them.

So, I will choose who and how many get to be in my inner circle. Being practical, I will select those who are like me so that there is less friction, less conflict. And I get to make up the rules regarding what areas of likeness I hold in the highest importance (economic status or values? sense of humor or taste in art? observed religion or sense of adventure?), and wanting to save time, I will quickly assess and assign labels and characteristics and behaviors to anyone or anything (who is not me) to assess this likeness.

By default, those who are not like me, and who do not reach the inner circle status, fall to a different category. A lower one obviously. By making my selections, I reinforce any already existing walls that prevent different groups from mingling much, or conversing much, or understanding one another much.Well, understanding one another as being members of a different group, that is, not necessarily as individuals. Because who really has time for that.

And so it goes.

At the root of all the isms and phobias in the world is one issue – the concept of exceptionalism. I first learned of this all-encompassing worldview from Carol Adams, author of *The Sexual Politics of Meat*, framer of eco-feminism scholarship, and tireless animal rights speaker. She is well-revered for many reasons – one of which is her ever growing bank of images that serve as examples from around the globe demonstrating the subjugation of women and women's bodies, the subjugation of non-White persons and bodies, and the subjugation of non-human animals and bodies. What all of these images have in common is that they illustrate exceptionalism. It's ok to subjugate women and members of the Global Majority and non-human animals because they are not me. The other is *less than* – period. Therefore, the rights and privileges and wealth that I get to enjoy and the rules and restrictions and poverty you have to endure are justified. It simply cannot be any plainer to understand – my group is better, or rather more exceptional, than your group; therefore, I get what and how much that I want. My view of you and my behavior toward you is the natural order of things.

When it comes to plants, other animals, and other people, this is a cynical 'take' on how we operate, but it holds truth. Perhaps we do not hold to this view individually, however we generally do as a species. When hard pressed to compare the value of different animals – a dog vs a goldfish perhaps or a hawk vs a centipede, the average person will default to a hierarchy instilled in us called the *Great Chain of Being*. Aristotle is credited for this philosophical hierarchy of the world where God is stationed at the top, followed by angels, humans, animals, plants and minerals. Within those categories, there are subcategories. Lions are more exceptional than foxes who are more exceptional than pigeons who are more exceptional than shrimp or grasshoppers, and so on. This bias is reinforced through religion, culture, holy day celebrations, the food we eat, the animals we allow (or don't) in our homes, the clothes we wear, the professions we choose, our friends, our partners, social media, anyone who wants to sell us anything...including the tourism industry.

It is because of this default toward otherness and toward exceptionalism that we must address our biases and the biases of our inner circle. Is it really that weird to hold our relationships with our companion animals to the same level as our human family members? Is it so hard to imagine that chickens or octopuses might also experience emotions? Does it make you odd to appreciate the virtues of non-human animals and of plants as much as or maybe even more than those of humans? And do you truly think that humans are more exceptional than the non-humans we share the earth with? I am not more in awe of my ability to speak a language that other humans can understand than the ability to live at 8000 meters under the sea like the Mariana hadal snailfish (*Pseudoliparis swirei*) or to fly at heights over 10,000 meters like the Ruppells Vulture (*Gyps rueppellii*) and Common Crane (*Grus grus*)? I for one laugh at the notion that humans are more exceptional than non-human animals because not one of us has the ability to navigate by echolocation like Microbats (*Chiroptera Microchiroptera*) or grow back our own limb like an Axolotl (*Ambystoma mexicanum*). By the way, we now know that some plants can communicate with each other via mycorrhizal fungi networks, chemical signals (Volatile Organic Compounds or VOCs) or electrical signals (Khait *et al.*, 2023).

This edited volume of chapters is a sledgehammer to the wall of exceptionalism. As Lucia Tomassini – our astute editor and guide through this collection – notes, tourism is embedded in othering. Tourism is mobility and how could you be mobile without running into someone, something, some place, some custom, some food, some language, some currency, some something that is different (read: less than) your own? We might glorify the other, we might vilify the other, we might call the other 'exotic' or 'interesting', but it is a rare and beautiful moment when we can catch the eye of another human who is very different from us on any number of characteristics and feel a warmth within us because for a few minutes, we are not othering. We are connecting. We are *being with* another. Or when we see a non-human animal and they see us and neither of us moves but we just consider each other for a moment and neither of us feels a need to chase or run or shoo or take a picture. That feels like equilibrium too.

The authors within this collection have begun what we must continue. They've shown us glimpses of equilibrium and *being with* another. And they've shown us what can happen if we don't. What is happening.

It's time for us to expand the concept of who can be in our inner circle. It's time to challenge the *Great Chain of Being*. It's time to laugh at ourselves when we catch ourselves thinking we are exceptional or better than some other person or species. It's time for this collection! Exceptionalism and othering is real and alive and dangerous. And it's also just quite ridiculous, isn't it?

Carol Kline
Appalachian State University, USA

Reference

Khait, I. *et al*. (2023) Sounds emitted by plants under stress are airborne and informative. *Cell* 186 (7), 1328–1336. https://doi.org/10.1016/j.cell.2023.03.009

1 Introduction – Entangled with the Others: What are the Implications for Tourism Futures?

Lucia Tomassini

> I'm not here to talk of futurology,
> but of literature [...] My confidence in the future
> of literature consists in the knowledge that there are things
> that only literature can give us, by means specific to it.
> I would, therefore, like to devote these lectures to certain
> values, qualities, or peculiarities of literature that are very
> close to my heart, trying to situate them
> within the perspective of the new millennium
> Italo Calvino, *Six Memos for the Next Millennium*, 1998

The Italian writer and journalist Italo Calvino, in his essay *Six Memos for the Next Millennium* (1988) emphasized the significance of reflecting on the values that characterize literature as it transitions to a new millennium, the current one. The essay originated from an invitation Calvino received in 1984 to deliver the Charles Eliot Norton poetry lectures at Harvard University the following year. However, these lectures never took place due to Calvino's sudden death. This edited book takes inspiration from Calvino's essay on the values and features helping to understand literature in the next millennium and it aims to shift such reflection to the tourism of this new millennium using features of the theory of space. It does so by focusing on the concepts of otherness and *othering* which play a key role in tourism, defined as 'a social, cultural, and economical phenomenon which entails the movement of people to countries or places outside their usual environment [...]' (UNWTO, n.d.). Such tourism-related 'movements of people' entail gazing at, encountering and experiencing territories (i.e. tourism destinations) and human and non-human actors (i.e. more-than-human host communities) understood and constructed as an otherness 'outside the usual environment' (Tomassini & Lamond, 2022).

Paraphrasing Calvino's words (1988: 1), this book is not here to talk of futurology, but of tourism with the confidence that the future of tourism consists in the knowledge that there are things that only tourism can give us, by means specific to it. Calvino's *Memos* (1988) were conceived as a set of future-oriented values crucial to the reading and interpretation of poetry and literature works from the first and second millennium (Botta & Re, 2023). Calvino's scholarly work embraces a multidisciplinary approach, addressing 'critical problems and themes relevant to the contemporary era and the future of the humanities' (1988: 2), aiming to unpack a 'multiplicity of the possible' (1988: 4). Taking Calvino's *Memos* (1988) as a starting point to explore this multiplicity and to indicate directions for the development of tourism in the new millennium, this edited book aims to present and question critical themes relevant to contemporary tourism and its futures. It does so through the lens of an eventful, relational and multidimensional space (Massey, 2005), characterized by posthuman entanglements between humans and non-humans, which are embedded in affirmative ethics (Braidotti, 2011, 2013), challenging notions of otherness and processes of *othering*.

When, how and why tourism has changed from 1946 to 2095 has been examined by Yeoman and McMahon-Beattie (2020) through historical and future turning points. This perspective article identifies mobility, Fordism, mass tourism and a modern-day leisure class as historical turning points and the political importance of tourism, footprint and transformational technologies as future turning points. As such, the political importance of tourism emerges from the debate around new mobilities paradigm (Sheller & Urry, 2006), the COVID-19 pandemic outbreak and its forced scattered immobility (Cheer *et al.*, 2021; Lew *et al.*, 2020), tourism and political and economic crises (Bianchi & Milano, 2024; Gibson, 2019), migration and social sustainability (Burrai *et al.*, 2022; Choe & Lugosi, 2022; Marcher *et al.*, 2019; Ormond & Vietti, 2021). In addition to the theme of non-human animals in tourism and justice for them (Bertella, 2021; Fennell, 2013, 2015; Fennell *et al.*, 2024; Kline, 2022). As such, the understanding of tourism – with its political importance, footprint and transformational technologies – has been interwoven with an ongoing critical reflection on its nature and role within a neoliberal, globalized world, underpinned by intense mobility of capital, goods, people and animals (Boluk *et al.*, 2019; Britton, 1980; Butcher, 2003; Tomassini & Lamond, 2022; Higgins-Desbiolles & Bigby, 2022). The COVID-19 pandemic crisis has renewed the attention for tourism's future and its redesign (Lew *et al.*, 2020), in addition to the existing critical reflections on spatial and mobility practices (Sheller, 2018; Sheller & Urry, 2006) and on the processes of *othering* embedded in chosen mobilities and forced displacements (Khazaal & Almiròn, 2021).

Within the context of the challenges posed by the Anthropocene epoch (Chakrabarty, 2017, 2018; Elhacham *et al.*, 2020; Ghosh, 2016), the waves of neo-nationalism emerging in Europe and worldwide (Brown, 2022; Zhou, 2022), and with the concept of space as a point of departure (Massey, 2005; Philo & Wilbert, 2000), this book aims to cast a novel light on the space of human and non-human others in tourism futures. Using Braidotti's words (2011: 25):

> [a]dvanced capitalism is a different engine in that it promotes the marketing of pluralistic differences and the commodification of the existence, the culture, the discourses of "others", for the purpose of consumerism. Consequently, the global system of the post-industrial world produces scattered and polycentered, profit-oriented power relations.

The development and growth of modern tourism and its mobility practices have grown within a Euclidean space that is scientifically measured and mapped by the Modern Western World (Tomassini & Lamond, 2022). The Eurocentric and anthropocentric philosophical approaches of Humanism and Enlightenment have legitimized leisure travelling – and its gaze (Urry, 1990) – as an interest towards an otherness conceived in terms of geographical distances, and ethnic, cultural and species diversities (Tomassini & Lamond, 2022). This has contributed to creating centers and peripheries (Wolf, 2010) as well as an anthropocentric approach to non-human animals and nature (Braidotti, 2011, 2013).

This edited book proposes a critical reflection on tourism and its futures by focusing on the concept of otherness and processes of *othering* with regard to the sociological and physical space of human and non-human others in tourism-related contexts. In doing so, it draws on future studies and theories to unpack and discuss emerging key themes in contemporary tourism, their interplay with future turning points in tourism development (Yeoman & McMahon-Bettie, 2020), and the ontological classifications of future studies as predictions, prognoses, science fiction and utopia/dystopia (Bergman *et al.*, 2010). The book's approach is grounded in Massey's (2005) conceptualization of space as a lively, heterogeneous and political response to global bio-capitalism, and the social inequalities generated by unbalanced power relations, where 'social' is not purely a human domain.

Outline of the Book

The contributions on the space of human and non-human actors in tourism futures presented in this book offer a set of critical reflections on various key aspects of tourism, where the concept of otherness and processes of *othering* emerge. In doing so, each contribution discusses the implications for tourism futures and possible scenarios. The book is divided into three parts and concludes with a chapter offering final reflections for the futures of tourism in this new millennium.

Part 1 – Theoretical Perspectives on Otherness, Processes of *Othering* and Space in Contemporary Tourism and its Future Development

The first part of this edited book offers a portfolio of contributions elaborating theoretical perspectives of the notion of otherness and processes of *othering* in the sociological space of tourism, discussing the interplay with future studies and theories.

Tomassini (Chapter 2) presents five of the six values identified by Calvino (1988) in his essay *Six Memos for the Next Millennium*, which discusses the qualities and legacy of literature as it transitions into a new millennium. Drawing inspiration from Calvino's essay, the chapter expands on the values of *lightness*, *quickness*, *exactitude*, *visibility* and *multiplicity*, applying them to tourism and future turning points in tourism development. Drawing on Massey's spatial theory and Braidotti's posthuman affirmative ethics, the chapter discusses Calvino's future-oriented values and their role in understanding the evolving landscape of tourism in this new millennium.

Burrai (Chapter 3) examines how and through which processes migration and tourism intersect in a posthuman space. By drawing on Sarah Ahmed's *ontology of stranger* and Rosi Braidotti's posthuman view on the concept otherness, the chapter aims to rethink tourism as a collective space where human/non-human encounters embrace the challenges linked to (forced) migration and respond to them mobilizing global and local actions. Drawing on future studies and theories, the chapter concludes envisioning two types of future scenarios, a utopian and a dystopian one.

Rizzolo and Fennell (Chapter 4) work towards a vision of justice for non-human animals that bridges otherness. Examining the case of captive elephant-based tourism in Asia, the chapter charts a path for ethical wildlife tourism between the danger of neocolonialism disregarding cultural variation and extreme cultural relativism failing to protect the well-being of non-human animals in tourism. Focusing on how elephant tourism futures might be informed by the intersection of the global and the local, resilience and diversification, the chapter presents potential elephant tourism scenarios and their different level of justice discussing the future(s) of ethical wildlife tourism.

Lapointe, Simard-Picard, Menu, Ethier and Laroque Labossière (Chapter 5) explore the complexities of the movement of deterritorialization and reterritorialization that contribute to the othering phenomenon of a more-than-human subject central to the tourism surrounding sugar time traditions in Québec, Canada – specifically, the maple tree. Drawing on assemblage theory, the chapter offers an analysis of assemblage tourism and maple syrup in Québec, examining how tourism practices contribute to processes of othering and reterritorialization. Drawing on future studies and poststructuralist approaches in critical future research, the chapter concludes by exploring the future of maple tourism through utopian and dystopian scenarios, highlighting its evolving entanglements with climate change, commodification and heritage.

Part 2 – Navigating Spatial Entanglements with the 'Other' and Implications for Tourism Futures

The second part of this edited book proposes a set of contributions on the spatial entanglement of human and non-human actors in tourism and their implications for tourism futures.

Bertella and Lombardini (Chapter 6) examine farm animals as the ultimate 'others' in rural tourism. By reviewing the scant tourism literature about farm animals and drawing on Keith Halfacree's rural space model, the chapter analyses animal–tourist encounters using the metaphor of scholars as tourists. In doing this, the chapter explores two scenarios – a dystopian and utopian one – involving alternative futures for the tourism and leisure scholarly literature in relation to farm animals.

Muldoon (Chapter 7) aims to shed light on how human and more-than-human animal lives are (de)humanized through processes of *othering* in popular discourses and poses critical questions regarding future mobilities in tourism and beyond in Europe. The chapter focuses on mobility discourses. Using critical discourse analysis, it examines the ways in which language is used to humanize rescued dogs while dehumanizing refugees, thereby justifying their (im)mobilities. The chapter concludes by offering dystopian and utopian future scenarios, helping prompt further reflection on alternative possible futures based on critical futures research.

Hoarau-Heemstra and Wally Ryan (Chapter 8) discuss the role of amateur underwater photographers in establishing a connection with the non-human otherness of marine life to understand it, connect with it and protect it, providing a foundation for the future regenerative tourism. Drawing from posthumanist perspectives, this chapter challenges traditional views, emphasizing interconnectedness and relationality and redefining agency through interconnected entanglements in the marine space. The chapter concludes by envisioning future utopian scenarios of regenerative diving destinations.

Äijälä, Haanpää and García-Rosell (Chapter 9) examine videography and the use of video as an essential component in both the research process and the dissemination of knowledge regarding human–animal encounters. By positing that animals and non-human subjects are valued by tourists and tourism organizations for their otherness and contribute to the relationality of tourism spaces, the chapter discusses how video methods can create new opportunities for co-creating spaces for non-human animals in the future of tourism. Drawing on future studies and critical futures research, the chapter elaborates on the role and potential of videography in envisioning utopian tourism futures.

Vietti (Chapter 10) reflects on the futures of tourism by interweaving anthropology and science fiction, highlighting the significant role of othering and encounters with the other in both fields. By examining the utopian attitude of the *solarpunk* science fiction genre – in contrast to the dystopian perspectives of *cyberpunk,* this chapter envisions a solarpunk tourism as a utopian future scenario capable of 'wandering' in multispecies urban environments, exploring the *twilight zone* where the real and the fantastic merge, and fostering relationships between humans and other-than-human creatures.

Part 3 – Views and Insights from Independent Researchers and Practitioners

The final part of this edited book collects a portfolio of views and insights from independent researchers and practitioners, bringing a multidisciplinary approach to tourism and the processes of othering embedded in its nature and development of the last decades. In doing so, these contributions present their experiences and points of view, reflecting on the implications for tourism futures.

Gainsforth (Chapter 11) offers a critical reflection on the interplay between tourism, new mobility patterns and urban processes of *othering,* with a focus on the Italian housing market in cultural cities. The chapter discusses the implications for the future of tourism and public policies aimed at ensuring sustainable tourism in the context of liveable cities and rural areas, driven by the principles of equity and social and spatial justice. The chapter concludes by offering a dystopian and utopian forecast with regard to tourism in Italian cultural cities.

Milovanovič (Chapter 12) presents the European project *Rural Migrantour – Paths of Recovery*, which focuses on rural territories to document, value and disseminate narratives related to minority cultures, border dynamics and contemporary migration issues. The project aims to empower young individuals in the field of responsible tourism by training them as cross-border guides capable of offering intercultural walks. The chapter concludes by envisioning inclusive cross-border cultural tourism as a utopian forecast for tourism futures.

Lovecky and Mueller (Chapter 13) present and discuss the potential of immersive media and the metaverse in Blue Tourism to bridge the gap between humans and the otherness of the marine environment and its non-human inhabitants, providing a means for deeper understanding and empathy. In doing so, the chapter examines two possible scenarios for the future of Blue Tourism in a metaverse-mediated world: a utopian scenario and a dystopian one.

Lenzi and Grasso (Chapter 14) introduce their Italian association, ETICOSCIENZA, which promotes discussions on the moral and ethical implications of anthrozoology and applied ethology. The chapter addresses their objective of mediating between ethics and science by adopting a biocentric and inclusive understanding of the human–animal relationship and encouraging critical reflections on processes of othering concerning animals and their perceived 'charisma'. In conclusion, a utopian future for animal-based tourism is presented.

Conclusive Reflections on Tourism Futures: Memos, Turning Points and Emerging Themes

This conclusive chapter synthesizes key insights from the book, which converge around a profound rethinking of tourism futures through the intersecting lenses of otherness, space and multispecies entanglements. What emerges collectively from the chapters is a call to envision tourism not merely as an economic or cultural phenomenon but as a dynamic and ethical space in which human and non-human actors co-construct futures. Each contribution brings to light how tourism encounters are embedded in relational complexities – between species, territories, technologies and power structures – demanding an epistemic shift toward posthuman and spatially conscious frameworks. By grounding the discussions in the spatial turn in social theory and posthuman ontology and ethics, the book constructs a critical scaffold for imagining tourism that is light in its footprint, quick in adaptive capacity, exact in ethical clarity, visible in its imaginaries, and multiple in its ontologies – in line with Italo Calvino's future-oriented values for this new millennium. As such, the final chapter synthesizes key insights from the book, exploring how tourism futures intersect with Calvino's memos, the spatial turn in the social sciences, the concept of otherness and processes of othering. Moreover, it examines how the chapters align with the three major turning points of tourism development – political and economic importance, tourism footprint and transformational technologies – highlighting their implications for sustainability, governance and emerging themes in contemporary tourism.

The collective contribution of the book lies in its insistence on embracing tourism's volatility as a space of opportunity – an open field where ethical and ecological responsibilities can be reimagined. Rather than offering prescriptive visions, the chapters advocate for a nuanced engagement with uncertainty, otherness, and more-than-human agency. This critical trajectory expands the contours of tourism futures, emphasizing the need for inclusive, responsive and theoretically grounded approaches capable of navigating the transformative challenges of the 21st century.

References

Bergman, A., Karlsson, J.C. and Axelsson, J. (2010) Truths claims and explanatory claims – an ontological typology of futures studies. *Futures* 42 (8), 857–865.

Bertella, G. (2021) *Wildlife Tourism Futures: Encounters with Wild, Captive and Artificial Animals*. Channel View Publications.

Bianchi, R. and Milano, C. (2024) Polycrisis and the metamorphosis of tourism capitalism. *Annals of Tourism Research* 104, 103731. https://doi.org/10.1016/j.annals.2024.103731

Boluk, K.A., Cavaliere, C.T. and Higgins-Desbiolles, F. (2019) A critical framework for interrogating the United Nations Sustainable Development Goals 2030 Agenda in tourism. *Journal of Sustainable Tourism* 27 (7), 847–864.

Botta, A. and Re, L. (2023) "Molteplicità del possible": Italo Calvino's Memos between the old and the new millennium. *California Italian Studies* 12 (1). https://dx.doi.org/10.5070/C312162261. Retrieved from https://escholarship.org/uc/item/51304840

Braidotti, R. (2011) *Nomadic Theory: The Portable Rosi Braidotti*. Columbia University Press.

Braidotti, R. (2013) *The Posthuman*. John Wiley & Sons.

Britton, S.G. (1980) The evolution of a colonial space-economy: The case of Fiji. *Journal of Historical Geography* 6 (3), 251–274.

Brown, G. (2022, November 15) *Nationalism is the Ideology of Our Age. No Wonder the World is in Crisis*. Retrived from https://www.theguardian.com/commentisfree/2022/nov/15/nationalism-world-crisis-rishi-sunak-joe-biden-g20

Burrai, E., Buda, D.M. and Stevenson, E. (2022) Tourism and refugees-crisis intersections: Co-creating tour guide experiences in Leeds, England. *Journal of Sustainable Tourism* 31 (12) 2680–2697.

Butcher, J. (2003) *The Moralization of Tourism: Sun, Sand...and Saving the World?* Routledge.

Calvino, I. (1988) *Six Memos for the Next Millennium*. Harvard University Press.

Chakrabarty, D. (2017) The politics of climate change is more than the politics of capitalism. *Theory, Culture & Society* 34 (2–3), 25–37.

Chakrabarty, D. (2018) Anthropocene time. *History and Theory* 57 (1), 5–32.

Cheer, J.M., Lapointe, D., Mostafanezhad, M. and Jamal, T. (2021) Global tourism in crisis: Conceptual frameworks for research and practice. *Journal of Tourism Futures* 7 (3), 278–294.

Choe, J. and Lugosi, P. (2022) Migration, tourism, and social sustainability. *Tourism Geographies* 24 (1), 1–8.

Elhacham, E., Ben-Uri, L., Grozovski, J., Bar-On, Y.M. and Milo, R. (2020) Global human-made mass exceeds all living biomass. *Nature* 588 (7838).

Fennell, D.A. (2013) Tourism and animal welfare. *Tourism Recreation Research* 38 (3), 325–340.

Fennell, D.A. (2015) Tourism and animal rights. *Tourism Recreation Research* 37 (2), 157–166.

Fennell, D.A., Kline, C., Mkono, M., Grimwood, B.S.R., Sheppard, V.A., Dashper, K., Rickly, J., Burns, G.L., Bertella, G., von Essen, E., García-Rosell, J.-C., Guo, Y., Hoarau-Heemstra, H., López López, Á., Quintero Venegas, G.J., Holladay, P.J., Cavaliere, C.T., Copeland, K., Danley, B. ... Madrid, R. (2024) Tourism, animals & the vacant niche: A scoping review and pedagogical agenda. *Current Issues in Tourism* 7 (22), 3820–3848. https://doi.org/10.1080/13683500.2023.2280704

Gibson, C. (2019) Critical tourism studies: New directions for volatile times. *Tourism Geographies* 23 (4), 659–677.

Ghosh, A. (2016) *The Great Derangement: Climate Change and the Unthinkable*. University of Chicago Press.

Higgins-Desbiolles, F. and Bigby, B.C. (2022) A local turn in tourism studies. *Annals of Tourism Research* 92, 103291.

Khazaal, N. and Almiròn, N. (2021) *Like an Animal: Critical Animal Studies Approaches to Borders, Displacement, and Othering*. Brill.

Kline, C. (2022) Socialing animal-based tourism. In F. Higgins-Desbiolles, A. Doering and C.B. Bigby (eds) *Socialising Tourism: Rethinking Tourism for Social and Ecological Justice* (pp. 195–213). Routledge.

Lew, A., Cheer, J.M., Haywood, M., Brouder, P. and Salazar, N.B. (2020) Visions of travel and tourism after the global COVID-19 transformation of 2020. *Tourism Geographies* 22 (3), 455–466.

Marcher, A., Kofler, I., Innerhofer, E. and Pechlaner, H. (2019) Perceptions of and interactions between locals, migrants, and tourists in South Tyrol. *Tourism Geographies* 24 (1), 56–72.

Massey, D. (2005) *For Space*. Sage.

Ormond, M. and Vietti, F. (2022) Beyond multicultural 'tolerance': Guided tours and guidebooks as transformive tools for civil learning. *Journal of Sustainable Tourism* 30 (2–3), 533–549.

Philo, C. and Wilbert, C. (2000) *Animal Spaces, Beasty Places: New Geographies of Human-Animal Relations*. Routledge.

Sheller, M. (2018) *Mobility Justice: The Politics of Movement in an Age of Extremes*. Verso Books.

Sheller, M. and Urry, J. (2006) The new mobilities paradigm. *Environment and Planning A: Economy and Space* 38 (2), 207–226.

Tomassini, L. and Lamond, I. (2022) Rethinking the space of tourism, its power-geometries, and spatial justice. *Journal of Sustainable Tourism* 31 (12), 2825–2838.

UNWTO (nd) *Glossary of Tourism Terms*. Retrieved from: https://www.unwto.org/glossary-tourism-terms

Urry, J. (1990) *The Tourist Gaze: Leisure and Travel in Contemporary Societies*. SAGE.

Yeoman, I. and McMahon-Beattie, U. (2020) Turning points' in tourism's development: 1946–2095, a perspective article. *Tourism Review* 75 (1), 86–90.

Zhou, S. (2022) The origins, characteristics and trends of neo-nationalism in the 21st century. *International Journal of Anthropology and Ethnology* 6 (18), 1–20.

Part 1

Theoretical Perspectives on Otherness, Processes of *Othering* and Space in Contemporary Tourism and Implications for its Future Development

2 The Future of Tourism: *Memos* for this New Millennium

Lucia Tomassini

Introduction

Calvino (1988), in his essay *Six Memos for the Next Millennium*, discusses the qualities and legacy of literature as it transitions into the new millennium. Calvino's *Memos* (1988) were conceived as a set of future-oriented values meant to guide the understanding of poetry as they transitioned from the first to the second millennium, prompting a critical discussion of issues and themes relevant to the contemporary era and the future of humanities (Botta & Re, 2023). In this chapter, Calvino's *Memos* (1988) are intertwined with future studies and theories (Yeoman & McMahon-Beattie, 2020a, 2020b) to identify key themes and issues of contemporary tourism and its futures.

Yeoman and McMahon-Beattie (2023) discuss how contemporary tourism operates in a world characterized by societal-scale risks such as climate change, pandemic outbreaks, war and artificial intelligence. As such, the future of tourism will unfold in a volatile, uncertain, complex and ambiguous world, where tourism futures research, awareness of possible future scenarios, as well as the capacity to foresee the directions of tourism futures will play a pivotal role (Yeoman & McMahon-Beattie, 2023). With regards to tourism development, Yeoman and McMahon-Beattie (2020b) identified six key turning points for the timeframe 1946–2095, which are historical and future turning points, meant as a guide from past to future for tourism development. The three historical turning points are mobility, Fordism and mass tourism, and a modern-day leisure class, while the three future turning points are the political importance of tourism, tourism footprint, and transformational technologies (Yeoman & McMahon-Beattie, 2020b).

With regards to the historical turning points of mobility, Fordism and mass tourism, and modern-day leisure class (Yeoman & McMahon-Bettie, 2020b), tourism – both as a cultural phenomenon and as an industry in Western cultures – originated in the last three centuries of the previous millennium. It evolved from the Grand Tour as an elitist aristocratic education and cultural experience (Brodsky-Porges, 1981) to Thomas Cook's working-class early mass tourism, and then the mass tourism of the late 20th century (Butcher, 2020). At the beginning of the 21st century, it further evolved with the rise of the sharing economy and digital platforms (Dolnicar, 2021; Hall *et al.*, 2022) as well as increased global mobility (Sheller, 2017, 2018) and temporary populations in urban areas (Brollo & Celata, 2022). Tourism and its mobility practices have become increasingly intertwined with the development of industrial capitalism, Fordism, and the advent of novel technologies, infrastructure, systems of production, and new topological and cartographic competencies (Borocz, 1992; Yeoman & McMahon-Beattie, 2020a). As such, the tourism industry has grown within a Euclidean space, scientifically measured, and mapped by the Modern Western World, where leisure, tourism and travelling were legitimized as an interest towards an 'otherness' made of geographical distances, ethnic, cultural and species diversities to gaze, experience and visit (Tomassini & Lamond, 2022).

With regards to the three future turning points of the growing political importance of tourism, tourism footprint and the role of transformative technologies (Yeoman & McMahon-Bettie, 2020), this chapter interweaves them with five of the Calvino's *Memos* (1988) – namely *Lightness, Quickness, Exactitude, Visibility* and *Multiplicity* to unpack and discuss key emerging themes of contemporary tourism and its future development (Aramberri, 2009; Gibson, 2019; Mura & Wijesinghe, 2021; Yeoman & McMahon-Beattie, 2023). The chapter discusses five out of six of Calvino's (1988) memos – i.e. future-oriented values – since the sixth one 'consistency' remained incomplete due to Calvino's death. In doing so, the chapter draws on Massey's (2005) ruminations on space and the spatial turn in the social sciences and the posthuman affirmative ethics of Braidotti (2011, 2013). As Massey (2005) stresses, space matters. It is not just a surface in which people, products and capital transit; 'easily this way of imagining space can lead us to conceive other places, people, and cultures simply as a phenomenon on "this" surface' (2005: 4). As such, this way of imagining space can lead to the commodification of 'others', which are understood as a naturalized or racialized 'otherness' (Braidotti, 2011) while posthuman affirmative ethics 'rests on an enlarged sense of interconnection between self and others, including the non-human or "earth" others, by removing the obstacles of self-centred individualism on the one hand, and the barrier of negativity on the other' (Braidotti, 2013: 190).

Drawing on the spatial turn of the social sciences and posthuman affirmative ethics, the chapter critically reflects on how Calvino's *Memos for the New Millennium* (1988) helps to identify key themes within the future turning points in tourism development (Yeoman & McMahon-Bettie, 2020) by discussing human and non-human otherness, processes of othering, and the role of space in contemporary tourism and its future development. Paraphrasing Calvino's words (1988: 1):

> *My confidence in the future of [tourism] consists in the knowledge that there are things that only [tourism] can give us, by means specific to it. I would therefore like to devote these [sections] to certain values, qualities, or peculiarities of [tourism] that are very close to my heart trying to situate them within the perspective of the new millennium.*

On Lightness

The future-oriented value of lightness, as a memo for this new millennium (Calvino, 1988), aligns with the future turning points of tourism's footprint and the role of transformational technologies in tourism development (Yeoman & McMahon-Bettie, 2020). Lightness has played a pivotal role in the final decades of the last millennium, characterized both by the ongoing dematerialization of economics, products and services and the transition to the weightless world of the digital economy (Coyle, 1998). This pursuit of lightness counterbalances the weight of the human-made mass – the 'anthropogenetic mass' (Elhacham *et al.*, 2020) – and the weight of our impacts on Earth: 'on average, for each person on the globe, anthropogenic mass equal to more than his or her bodyweight is produced every week' (2020: 442). In this context, the search for lightness aligns with the goals of sustainability science, which aim to achieve positive social and environmental transformations by minimizing or eliminating negative human impacts on humans (Aktaş, 2022). Connecting the value of lightness to tourism encourages further reflection on the legacy and future of tourism, particularly its sustainability, as it transitions into the new millennium.

While travelling implies the burden of the tangible and intangible belongings we bring with us (e.g. luggage and cultural background), such a burden has been dematerializing in the last century and a few decades. Our travelling has largely become lighter due to the emergence of a modernist portability culture (Ridge, 2017) as well as digital technologies, accessible networked global services, peer-to-peer services, sharing, and the growing desire for an ease of movement and 'travelling light' (Ridge, 2017). Moreover, contemporary tourism mobilities build on the idea – and myth – of global citizenship (Caton *et al.*, 2014), resonating with the tourist syndrome (Franklin, 2003) of our contemporary society and lives made of temporary-limited experiences and divided by the freedom to move (Bauman, 1996). Yet, our desire to travel light across different destinations carries the burden of an industry caught between the lightness and dematerialization of tourism experiences and the heavy impacts – that is, footprint – the tourism industry has on the sociological and physical spaces of human and non-human host communities.

As a memo for the tourism of this new millennium, lightness can be understood through the lens of a relational sustainability (West *et al.*, 2020, 2024) and the concept of metamorphosis, echoing Ovid's *Metamorphoses* (2004). On the one hand, the relational approach to sustainability is a paradigm examining sustainability while advocating the importance of interconnectedness, reciprocity and mutuality between humans and the nonhuman world (West *et al.*, 2020, 2024). As such, it aligns with the posthuman critical theory discussing widespread belonging, nomadic subjectivity and affirmative ethics rooted in becoming intimate with otherness and diversity (Braidotti, 2011, 2013). On the other hand, Ovid's (2004) *Metamorphoses* celebrates transformation and change as a unifying process across time that bridges otherness and diversity while disclosing wonders (Wheeler, 1999) and wandering viewpoints (Iser, 1978, 2009).

The relational approach to sustainability questions the anthropocentrism of the definition of sustainable development of *Our Common Future* (Brundtland, 1987), the horizon of sustainable development goals (Boluk *et al.*, 2019), as well as the misleading concept of sustainable growth (Monbiot, 2012). Hence, the notion of relational sustainability – combined with metamorphoses prompting ongoing regenerative transformations – helps envision the lightness of the tourism of this new millennium. Such lightness is here understood as the capacity to embrace different perspectives and enact multiple affective belongings within more-than-human communities in the more-than-human spaces of tourism destinations. Lightness is not a lack of commitment but – on the contrary – it is a 'travelling light' through de-materialized experiences rooted in a sense of accountability, empathy and intimacy with otherness and diversity. Being the 'other' is, indeed, a temporary and spatially localized status while we move through wandering viewpoints (Iser, 1978; Wheeler, 1999) of human and/or non-human travellers and guests.

On Quickness

The future-oriented value of quickness, as a memo for this new millennium (Calvino, 1988), aligns with the future turning point of the role of transformational technologies in tourism development (Yeoman & McMahon-Beattie, 2020b). The value of quickness can be better understood in comparison to its opposite, slowness, as quickness is not a value per se, but rather a quality we praise to gain more time to use – or to lose (Calvino, 1998). Since the last few decades of the previous millennium, the transition to a digital economy has been characterized by quickness, through instant access to information, services, real-time web navigation as well as to a hyper connectivity via digital identities and social media (Coyle, 1998). Quickness, therefore, can be viewed as a key quality of contemporary society and digital economy, as well as a value historically intertwined with the technological and industrial developments of the modern era in which faster means of production and transport have conveyed efficiency and performance, often over human well-being (Bruni, 2020).

Quickness has played a central role in the critical understanding of the societal transformations and economic changes of the last century. For instance, the dehumanized working pace of the industrial production lines pictured in Charlie Chaplin's *Modern Times* (1936) as well as the celebration of speed and technological progress of the artistic avant-garde movement *Futurismo* in the early 20th century (Ialongo, 2016). In 'Speed Kills – Fast is never fast enough', Mark Taylor (2014, October 20) posits: '[...] with the emergence of industrial capitalism, the primary values governing life became work, efficiency, utility, productivity, and competition. When Frederick Winslow Taylor took his stopwatch to the factory floor in the early 20th century to increase workers' efficiency, he began a high-speed culture of surveillance so memorably depicted in Charlie Chaplin's *Modern Times* [...]'. Hence, connecting quickness to tourism entails reflecting on how speed, efficiency and instant access to information can shape tourism in this new millennium, acknowledging travellers' desire to move with agility, precision and speed without losing the depth of the travel experience.

In tourism studies, quickness has been largely investigated through the lenses of performance, productivity and efficiency (Marrocu & Paci, 2011; Sainaghi *et al.*, 2017). Yet, critical tourism studies have been questioning this growth-oriented perspective advocating the paradigm shift towards degrowth for an emerging agenda for tourism research and praxis (Boluk *et al.*, 2019; Fletcher *et al.*, 2019). Degrowth has been discussed as a slowing down by decelerating our Western societies and economies by design (Kallis *et al.*, 2020; Saito, 2024). Hence, this chapter examines quickness as a memo for this new millennium through the classic adage and oxymoron *Festina Lente*. Originally attributed to the Roman Emperor Augustus in the 1st century, used by the Italian Medici family in the 16th century, the adage *Festina Lente* is presented by Calvino (1988: 48) as 'hurry slowly' and his personal motto: 'my work as a writer has from the beginning aimed at tracing the lightning flashes of the mental circuits that capture and link points distant from each other in space and time [...] I have always aimed at the image and the motion that arises naturally from the image, while still being aware that one cannot speak of a literary result until this stream of imagination has been turned into words' (1988: 48).

Envisioning quickness as *Festina Lente* – 'make haste slowly' – allows us to think about the future of tourism as a balance between momentum and mindfulness, mediated and co-created via an increased use of transformational technologies (Yeoman & McMahon-Beattie, 2020b). This entails travelling with awareness, moderation and precision in destination choice, tourism mobility practices and encounters while translating the travel's idea into a real journey. Hence, the paradox of balancing speed with caution, as a memo for this new millennium, means moving through the space of human and non-human others as hurrying slowly, through careful and attentive deceleration preventing processes of *othering*. This involves utilizing the hyper-connectivity of our era, as well as rapid access to information, networks and mobilities as opportunities to engage with and empathize with the needs of the host human and non-human community inhabiting the local space of tourism destinations (Higgins-Desbiolles & Bigby, 2022; Tomassini & Cavagnaro, 2020). Quickness as *Festina Lente* also means being aware of – and acting accordingly to – tourism's carbon emissions (Gössling *et al.*, 2023), the interplay between tourism, mobility (un)justice and politics of movement (Sheller, 2018), as well as embracing an affective posthuman approach to a cosmopolitan hospitality for a more just tourism (Guia & Jamal, 2023).

On Exactitude

The future-oriented value of lightness, as a memo for this new millennium (Calvino, 1988), aligns with the future turning points of tourism's footprint and the role of transformational technologies in tourism development (Yeoman & McMahon-Beattie, 2020b). Calvino (1988) conceptualizes exactitude as a crucial literary and intellectual principle, emphasizing precision, clarity and a well-structured, effective form that opposes vagueness and confusion. He describes exactitude as having three key dimensions: (1) the creation of a well-defined and evocative image; (2) the use of precise language that avoids redundancy and ambiguity; and (3) the development of a structured and coherent narrative capable of conveying complexity without disorder. This notion of exactitude transcends literature, serving as an antidote to superficiality and chaos, encouraging a deeper commitment to language, ideas and space (Massey, 2005), and the bridge between theory and practice in shaping contemporary realities and their space.

As contemporary tourism unfolds in a volatile, uncertain, complex and ambiguous world – characterized by societal-scale risks such as climate change, pandemic outbreaks, geopolitical instability and artificial intelligence (Yeoman & McMahon-Beattie, 2023) – exactitude emerges as a critical principle for ensuring tourism develops in a meaningful, sustainable and adaptable manner. The application of precision in tourism planning, communication and experience design allows for structured foresight, enabling destinations to implement policies that balance economic benefits with environmental and cultural sustainability (Gössling *et al.*, 2009). As such, exactitude can play a key role in destination management, resource allocation, and the mitigation of tourism impacts such as urban touristification, gentrification of historical centers, carbon emissions from travel and hospitality, and the erosion of local cultural identities (Seraphin *et al.*, 2018).

In the realm of tourism branding and storytelling, exactitude is equally crucial. In an era of information overload, destinations must craft compelling yet truthful narratives that highlight their unique characteristics while preventing cultural misrepresentation and unsustainable visitor influx. As Urry (1990) posits in *The Tourist Gaze*, tourism experiences are shaped by the way destinations are framed and marketed; travellers often seek experiences that align with their preconceived notions and idealized representations. However, when marketing and communication strategies prioritize broad appeal and economic growth over accuracy and host community wellbeing, destinations risk fostering unrealistic expectations, leading to visitor dissatisfaction and cultural degradation. A case in point is the over-commercialization of Venice, where mass tourism has diluted local authenticity and contributed to the displacement of residents (Russo & Scarnato, 2018). By adhering to Calvino's principle of exactitude, tourism planning and storytelling should remain both engaging, inclusive and accurate, fostering respectful interactions between tourists and host communities in accordance with the principles of posthuman affirmative ethics (Braidotti, 2013).

The rise of data-driven and smart tourism illustrates the increasing relevance of exactitude in the industry. Technologies such as artificial intelligence, big data analytics and predictive modelling enable businesses and destinations to offer highly personalized, efficient and environmentally conscious services (Gretzel *et al.*, 2015). These data-driven approaches align with Calvino's emphasis on structured and precise planning, ensuring that tourism remains responsive to both consumer demands and host community needs. Exactitude also plays a fundamental role in tourism futures research. Moreover, as Yeoman (2012) highlights, the uncertain nature of global travel necessitates structured, evidence-based forecasting and scenario planning. As such, data-driven forecasting and scenario planning approaches embody Calvino's call for structure and precision by developing alternative futures based on emerging trends and data-driven insights.

In sum, Calvino's (1988) concept of exactitude challenges the tourism industry to adopt precision in its policies, narratives and technological advancements. By embracing structured, thoughtful approaches, tourism can navigate future uncertainties while maintaining clarity and purpose. As destinations face pressing global challenges, exactitude serves as a valuable guiding principle for fostering meaningful, sustainable and well-managed tourism experiences.

On Visibility

The future-oriented value of visibility, as a memo for this new millennium (Calvino, 1988), aligns with the future turning points of the growing political and economic importance of tourism and the role of transformational technologies in tourism development (Yeoman & McMahon-Beattie, 2020b).

Visibility in Calvino's essay (1988) refers to the power of imagination and visualization: 'If I have included visibility in my list of values to be saved, it is to give warning of the danger we run in losing a basic human faculty: the power of bringing visions into focus with our eyes shut [...] and in fact of thinking in terms of images' (1988: 92). Abstract thinking, creativity and visual imagination allow for the creation of parallel visions, realities and scenarios. Hence, visibility – by exercising high fantasy and the loftier part of imagination – helps in discussing the space of human and non-human others in tourism futures with fantasy, imaginative culture and creative thinking. Ghosh's words clarify the connection between culture and imagination: 'the climate crisis is also a crisis of culture and thus of the imagination' (2016a: 9). The number of imaginative works and fictional narratives able to describe the contemporary societal and environmental challenges – like the climate crisis – is surprisingly scant (Gosh, 2016b). Yet, visual thinking and imaginative culture have been playing a pivotal role in human history and cultural and artistical evolution (Carroll *et al.*, 2020) as well as in the visual culture of western societies (Jenks, 1995). Drawing on the concept of the tourist gaze (Urry, 1990), this section examines the value of visibility in tourism futures through the notion of *imagineering* – combining 'imagination' and 'engineering'. The term *imagineering* was popularized by Walt Disney in the 1950s (Deherer, 2011) and then it has been permeating various fields as events design strategy (Ouwens, 2016), city futures (Ratcliff & Kraeczyk, 2011) and – still in an embryonal way – tourism studies (Salazar, 2013).

Drawing on gaze theory and the Foucauldian gaze as an instrument of power and control (Foucault, 1976), the tourist gaze (Urry, 1990) argues how the prominence of visual elements in modern culture is reflected in tourism. In doing so, the tourist gaze posits that our motivations to visit destinations and the ways we learn to visually enjoy them are shaped by social structures rather than being purely personal or independent (Caletrío, 2016). Thus, changes in tourism are inherently connected to broader societal changes as visual imagination and visual thinking are embedded in the desire to travel and visually experience geographical, cultural or species otherness as well as to engage with it. In tourism studies, the gaze has been largely explored (Samarathunga & Cheng, 2020), from the tourist gaze (Urry, 1990), the host gaze (Moufakkir, 2011), the mutual gaze (Maoz, 2006), till the (missing) animal gaze (Tomassini & Bertella, 2023). In the transition of tourism into this new millennium – and with the aim of reflecting on the space of human and non-human others in tourism futures – the value of visibility is here discussed through the concept of *imagineering*. Drawing on Walt Disney's approach, *imagineering* refers to the creative process of designing and building immersive, story-driven experiences, blending artistry, technology and innovation (Dehrer, 2011). Such concept now applies broadly not only to creative industries involving visionary thinking and practical execution.

Envisioning the value – and legacy – of visibility in tourism futures as *imagineering* allows for enacting a disruptive, creative and visual thinking, as well as a novel imaginative culture grounded in posthuman affirmative ethics and nomadic subjectivity (Braidotti, 2011, 2013). As such, the value of *imagineering* can play a foundational role for posthuman tourism research (Guia & Jamal, 2020) and justice (Guia, 2021). Enacting *imagineering* in the tourism of this new millennium – by combining imagination, engineering and technologies – challenges socially constructed gazes, creates a more-than-human tourism space free from unbalanced power relations (Tomassini & Lamond, 2022) and stereotypical western-centered imaginaries (*Let's Save Africa – Gone wrong*, n.d.), as well as challenging anthropocentric dominant discourses (Tomassini, 2024). Visibility as *imagineering* entails adopting a periscoping attitude to shed light on niche research areas – for example, the animal gaze in tourism studies (Tomassini & Bertella, 2023) – or potentialities embedded in novel fields of study – for example immersive technology for tourism experiences (Fan *et al.*, 2022; Pratisto *et al.*, 2022) – that may be overlooked or constrained by traditional research methodologies and conservative areas of interest (Hiemstra, 2017).

On Multiplicity

The future-oriented value of multiplicity, as a memo for this new millennium (Calvino, 1988), aligns with the future turning points of the growing political and economic importance of tourism and its footprint (Yeoman & McMahon-Beattie, 2020b). The value of multiplicity – to some extent – subsumes the previous ones of *lightness*, *quickness* and *visibility* in the attempt to look at the legacy and guiding qualities of tourism as it transits into this new millennium. Moreover, multiplicity is a driving concept of the theoretical lenses informing this chapter: posthumanism and its nomadic subjectivity and relational ontology (Braidotti, 2013) and Massey's (2005) spatial theory and its propositions for a space that is relational, constituted through interaction, a sphere of multiplicity and coexisting heterogeneity, and always under construction. In Calvino's essay (1988) multiplicity is discussed with regards to the encyclopedic nature – 'encyclopedism' – of the literature of the 20th century aiming to be a method of knowledge and a network of connections able to hold multiple perspectives and weave together complex narratives. To some extent, the value of multiplicity in the tourism studies of the 20th century largely aligned with this encyclopaedic approach viewing tourism as a field of study combining different disciplines and perspectives, as well as a method to build knowledge and networks of connections among different actors. In the last two decades, such an 'encyclopedic' approach to tourism studies became evident since several academic publications appeared as encyclopaedias of tourism. To name a few of them: *Encyclopaedia of*

Tourism Management and Marketing (Buhalis, 2022), *Encyclopedia of Sustainable Tourism* (Cater *et al.*, 2015), *Encyclopedia of Tourism* (Jafari, 2001).

The etymological root of the word 'encyclopedia' means – from the ancient Greek '*enkyklios paideia*' (*ἐγκύκλιος παιδεία*) – the search of a 'circular instruction' created through a set of doctrines forming a comprehensive education (Dionigi, 2019). Nevertheless, such 'attempt to exhaust knowledge of the world by enclosing it in a circle' (Calvino, 1988: 116) seems to no longer match this new millennium, since 'we can no longer think in terms of a totality that is not potential, conjectural, and manifold' (1988: 116). Thus, this section questions any 'encyclopedism' by advocating the value of multiplicity in tourism futures as an open, never given, spatiality, made of a multiplicity of connections and relations among human and non-human actors (Massey, 2005; Tomassini & Lamond, 2022). As such, tourism in this new millennium should take place in a space that is not just a surface to transit but is rather an eventful relational open space because: 'for the future to be open, space must be open too' (Massey, 2005: 12). This means a tourism space open to the diversity and variety of the human and non-human actors who inhabit and move through such space. Consequently, this open multiple relational space questions the idea of a diminished otherness – e.g. racial, cultural and species – as well as any processes of *othering*, valuing instead a relational ontology made of a widespread belonging (Braidotti, 2011, 2013).

Envisioning the value of multiplicity in the space of human and non-human actors in the tourism of this new millennium calls for a more radical approach, one that unpacks the potential embedded in tourism's temporally limited nature and its willingness to discover novel experiences and dimensions. The value of multiplicity understood as an open relational space, aligns with posthuman affirmative ethics (Braidotti, 2013), which is committed to engaging meaningfully with the complexity of different cultures, stories, points of view, gazes, discourses and experiences of the human and non-human communities inhabiting the tourism destination. Thus, multiplicity in tourism futures would challenge the traditional consumptive and commodifying nature of the tourism industry, rethinking it through critical lenses that deconstruct tourism beyond the host and guest dichotomy and challenge the tourism destination as an anthropocentric tourist space, often disregarding the desires, needs and interests of local human and non-human communities (Higgins-Desbiolles & Bigby, 2022; Tomassini & Cavagnaro, 2020). Hence, the tourism of this new millennium should be grounded in the awareness that our lives are made of temporally limited, localized personal and professional experiences, shaped by living, working and experiencing different cities, territories and countries as members of a cosmopolitan, more-than-human society.

Conclusions

This chapter has explored the intersection of Italo Calvino's *Six Memos for the Next Millennium* (1988) with future studies in tourism, linking five of Calvino's literary values –*lightness, quickness, exactitude, visibility* and *multiplicity* – to the three future turning points identified by Yeoman and McMahon-Beattie (2020b): the growing political and economic importance of tourism, tourism footprint and the role of transformational technologies. Through this interdisciplinary dialogue, the chapter has demonstrated how Calvino's memos offer a conceptual framework for critically analyzing tourism's transition into the new millennium. Drawing on the spatial turn of the social sciences (Massey, 2005) and posthuman affirmative ethics (Braidotti, 2013) this chapter offers theoretical and practical insights for unpacking and questioning human and non-human otherness and processes of *othering* in contemporary tourism and its future development.

The value of *lightness* aligns with the future turning point of tourism footprint and transformational technologies. As tourism development becomes increasingly entangled with environmental concerns, *lightness* urges a move toward dematerialized experiences, sustainable travel models and relational sustainability. The weight of tourism's ecological and sociocultural impacts, ranging from CO_2 emissions to gentrification, necessitates a shift toward more mindful and responsible tourism practices co-created and mediated by transformational technologies. A tourism future guided by *lightness* embraces ethical, immersive and low-impact approaches, reducing its footprint while fostering deeper engagement with destinations.

Quickness, in turn, resonates with the role of transformational technologies in tourism. While digitalization, artificial intelligence and smart tourism promise efficiency, connectivity and hyper-mobility, the value of *quickness* also warns against unchecked acceleration. Balancing speed with mindfulness – *festina lente* (hurry slowly) – emerges as a key principle for tourism futures, ensuring that technological progress enhances rather than diminishes meaningful travel experiences. *Quickness*, when strategically employed, can drive innovation without compromising sustainability or wellbeing.

The future-oriented value of *exactitude* connects with the future turning point of tourism footprint and the role of transformational technologies, particularly in the realm of destination management, planning and impact mitigation. As tourism faces increasing pressures from over-tourism, cultural commodification and environmental degradation, exactitude calls for clarity, precision and well-structured policies. From carbon reduction strategies to equitable visitor distribution, an exacting approach ensures that tourism operates within defined sustainability parameters, preventing excessive strain on host communities and ecosystems.

Visibility is closely linked to the growing political and economic importance of tourism and the role of transformational technologies. In a world where real and digital representation and perception shape tourism development, *visibility* highlights the power of imagination, storytelling, digitally mediated experiences, and both tourist and host gazes in framing destinations. However, visibility must be approached critically – tourism narratives should not only attract but also challenge dominant perspectives, ensuring that tourism development respects and authentically represents diverse cultural identities. *Imagineering*, as discussed in this chapter, becomes a vital tool in shaping ethical, inclusive – and technologically mediated – future tourism experiences.

Finally, *multiplicity* corresponds to the future turning point of tourism's growing political role and its footprint, emphasizing the need for tourism governance models that embrace diversity, complexity and intersectionality. Rather than viewing tourism as a singular economic force, multiplicity envisions tourism futures as open, relational spaces where human and non-human actors coexist. This perspective aligns with posthuman affirmative ethics, recognizing the interconnectedness of tourism with broader societal, ecological and political dynamics.

By weaving together Calvino's literary insights with tourism futures research, this chapter underscores the importance of cross-disciplinary thinking in addressing the challenges of tourism in the new millennium. In an era of uncertainty and transformation, Calvino's memos provide a framework for reimagining tourism as a dynamic, adaptive and ethically engaged practice – one that navigates the complexities of the future with clarity, creativity and responsibility.

References

Aktaş, C.B. (2022) Dematerialization: Needs and challenges. In W. Leal Filho, A.M. Azul, F. Doni and A.L. Salvia (eds) *Handbook of Sustainability Science in the Future*. Springer.

Aramberri, J. (2009) The future of tourism and globalization: Some critical remarks. *Futures* 41 (6), 367–376.

Bauman, Z. (1996) *Tourists and Vagabonds: Heroes and Victims of Postmodernity*. Institut für Höhere Studien (IHS), Wien. https://nbn-resolving.org/urn:nbn:de:0168-ssoar-26687

Boluk, K.A., Cavaliere, C.T. and Higgins-Desbiolles, F. (2019) A critical framework for interrogating the United Nations Sustainable Development Goals 2030 Agenda in tourism. *Journal of Sustainable Tourism* 27 (7), 847–864.

Borocz, J. (1992) Travel-capitalism: The structure of Europe and the advent of the tourist. *Comparative Studies in Society and History* 34 (4), 708–774.

Botta, A. and Re, L. (2023) "Molteplicità del possible": Italo calvino's memos between the old and the new millennium. *California Italian Studies* 12 (1). http://dx.doi.org/10.5070/C312162261

Braidotti, R. (2011) *Nomadic Theory: The Portable Rosi Braidotti*. Columbia University Press.

Braidotti, R. (2013) *The Posthuman*. John Wiley & Sons.

Brodsky-Porges, E. (1981) The grand tour travel as an educational device 1600–1800. *Annals of Tourism Research* 8 (2), 171–186.

Brollo, B. and Celata, F. (2023) Temporary populations and sociospatial polarisation in the short-term city. *Urban Studies* 60 (10), 1815–1832.
Brundtland, G.H. (1987) *Our Common Future: Report of the World Commission on Environment and Development.* Oxford University Press.
Bruni, L. (2020) *Economics, Wellbeing and Happiness: A Historical Perspective.* Edward Elgar Publishing.
Buhalis, D. (2022) *Encyclopedia of Tourism Management and Marketing.* Edward Elgar Publishing.
Butcher, J. (2020) The development of mass tourism. In I. Yeoman and U. McMahon-Beattie (eds) *The Future Past of Tourism: Historical Perspectives and Future Evolutions* (pp. 20–35). Channel View Publications.
Caletrío, J. (2019, March 25) The tourist gaze by John Urry, *Mobile Lives Forum.* https://forumviesmobiles.org/en/node/12906/printable/print
Calvino, I. (1988) *Six Memos for the Next Millennium.* Harvard University Press.
Carroll, J., Clasen, M. and Jonsson, E. (2020) *Evolutionary Perspectives on Imaginative Culture.* Springer.
Cater, C., Garrod, B. and Low, T. (2015) *The Encyclopedia of Sustainable Tourism.* CABI.
Caton, K., Schott, C. and Daniele, R. (2014) Tourism's imperative for global citizenship. *Journal of Teaching in Travel & Tourism* 14 (2), 123–128.
Chaplin, C. (1936) *Modern Times.* United Artists.
Coyle, D. (1998) *The Weightless World: Strategies for Managing the Digital Economy.* MIT Press.
Dionigi, I. (2019) *Osa Sapere. Contro La Paura e l'ignoranza.* Solferino.
Dolnicar, S. (2021) Sharing economy and peer-to-peer accommodation – a perspective paper. *Tourism Review* 76 (1), 34–37.
Dehrer, G. (2011) Imagineers in search of the future. *The Futurist* 45 (2), 36.
Elhacham, E., Ben-Uri, L., Grozovski, J., Bar-On, Y.M. and Milo, R. (2020) Global human-made mass exceeds all living biomass. *Nature* 588 (7838), 442–444.
Fan, X., Jiang, X. and Deng, N. (2022) Immersive technology: A meta-analysis of augmented/virtual reality applications and their impact on tourism experience. *Tourism Management* 91, 104534.
Fletcher, R., Murray Mas, I., Blanco-Romero, A. and Blázquez-Salom, M. (2019) Tourism and degrowth: An emerging agenda for research and praxis. *Journal of Sustainable Tourism* 27 (12), 1745–1763.
Foucault, M. (1976) *The Birth of the Clinic: An Archaeology of Medical Perception.* Vintage Books.
Franklin, A. (2003) The tourist syndrome: An interview with Zygmunt Bauman. *Tourist Studies* 3 (2), 205–217.
Gibson, C. (2019) Critical tourism studies: New directions for volatile times. *Tourism Geographies* 23 (4), 659–677.
Ghosh, A. (2016a) *The Great Derangement: Climate Change and the Unthinkable.* University of Chicago Press.
Gosh, A. (2016b, October 28) Amitav Gosh: Where is the fiction about climate change? The Guardian. https://www.theguardian.com/books/2016/oct/28/amitav-ghosh-where-is-the-fiction-about-climate-change-
Gössling, S., Hall, C.M. and Weaver, D.B. (2009) *Sustainable Tourism Futures: Perspectives on Systems, Restructuring and Innovation.* Routledge.
Gretzel, U., Sigala, M., Xiang, Z. and Koo, C. (2015) Smart tourism: Foundations and developments. *Electron Markets* 25, 179–188.
Gössling, S., Balas, M., Mayer, M. and Sun, Y.-Y. (2023) A review of tourism and climate change mitigation: The scales, scopes, stakeholders and strategies of carbon management. *Tourism Management* 95, 104681.
Guia, J. (2021) Conceptualizing justice tourism and the promise of posthumanism. *Journal of Sustainable Tourism* 29 (2–3), 503–519.

Guia, J. and Jamal, T. (2020) A (Deleuzian) posthumanist paradigm for tourism research. *Annals of Tourism Research* 84, 102982.
Guia, J. and Jamal, T. (2023) An affective and posthumanist cosmopolitan hospitality. *Annals of Tourism Research* 10, 103569.
Hall, C.M., Prayag, G., Safonov, A., Coles, T., Gössling, S. and Naderi Koupaei, S. (2022) Airbnb and the sharing economy. *Current Issues in Tourism* 25 (19), 3057–3067.
Hiemstra, N. (2017) Periscoping as a feminist methodological approach for researching the seemingly hidden. *The Professional Geographer* 69 (2), 329–336.
Higgins-Desbiolles, F. and Bigby, B.C. (2022) A local turn in tourism studies. *Annals of Tourism Research* 92, 103291.
Ialongo, E. (2016) Futurism from foundation to world war: The art and politics of an avant-garde movement. *Journal of Modern Italian Studies* 21 (2), 306–323.
Iser, W. (1978) *The Act of Reading: A Theory of Aesthetic Response.* Johns Hopkins University Press.
Iser, W. (2009) Texts and readers. *Discourse Processes* 3 (4), 327–343.
Jafari, J. (2001) *The Encyclopedia of Tourism.* Routledge.
Jenks, C. (1995) *Visual Culture.* Routledge.
Kallis, G., Paulson, S., D'Alisa, G. and Demaria, F. (2020) *The Case for Degrowth.* Wiley & Sons.
Maoz, D. (2006) The mutual gaze. *Annals of Tourism Research* 33 (1), 221–239.
Marrocu, E. and Paci, R. (2011) They arrive with new information: Tourism flows and production efficiency in the European regions. *Tourism Management* 32 (4), 750–758.
Massey, D. (2005) *For Space.* Sage.
Monbiot, G. (2012, June 22) How sustainability became sustained growth. *Monbiot.com.* https://www.monbiot.com/2012/06/22/how-sustainability-became-sustained-growth/
Moufakkir, O. (2011) The role of cultural distance in mediating the host gaze. *Tourist Studies* 11 (1), 73–89.
Mura, P. and Wijesinghe, S.N.R. (2021). Critical theories in tourism – a systematic literature review. *Tourism Geographies* 25 (2–3), 487–507.
Ouwens, F. (2016) The role of imagineering in event design strategy for the business event industry. In L.A.T. van der Woude, H.J.S.G. Meijer and A. M. L.J. Dijkstra (eds) *The Business of Events: The Role of Strategy in the Event Industry* (pp. 53–66). Routledge.
Ovid (2004) *Metamorphoses (D. Raeburn, Trans.).* Penguin Classics.
Pratisto, E.H., Thompson, N. and Potdar, V. (2022) Immersive technologies for tourism: A systematic review. *Information Technology & Tourism* 24, 181–219.
Ratcliffe, J. and Kraeczyk, E. (2011) Imagineering city futures: The use of prospective through scenarios in urban planning. *Futures* 43 (7), 642–653.
Ridge, E. (2017) *Portable Modernisms: The Art of Travelling Light.* Edinburgh University Press.
Russo, A.P. and Scarnato, A. (2018) 'Barcelona in common': A new urban regime for the 21st-century tourist city? *Journal of Urban Affairs* 40 (4), 455–474.
Saito, K. (2024) *Slow Down: A Manifesto for a Sustainable Future.* Astra House.
Sainaghi, R., Phillips, P. and Zavarrone, E. (2017) Performance measurement in tourism firms: A content analytical meta-approach. *Tourism Management* 59, 36–56.
Salazar, N.B. (2013) Imagineering otherness: Anthropological legacies in contemporary tourism. *Anthropological Quarterly* 86 (3), 669–696.
Samarathunga, W.H.M.S. and Cheng, L. (2020) Tourist gaze and beyond: State of the art. *Tourism Review* 76 (2), 344–357.
Seraphin, H., Gowreesunkar, V., Ambaye, M. and Bonnardel, V. (2018) Over-tourism and the fall of Venice as a destination. *Journal of Destination Marketing & Management* 9, 374–376.
Sheller, M. (2017) From spatial turn to mobilities turn. *Current Sociology* 65 (4), 623–639.

Sheller, M. (2018) *Mobility Justice: The Politics of Movement in an Age of Extremes.* Verso Books.
Taylor, M. (2014, October 20) *Speeds Kills. Fast is Never Fast Enough.* https://www.chronicle.com/article/speed-kills
Tomassini, L. (2024) Posthumanism and animal geography: The case of Freya the walrus on media. *Tourism Geographies* 26 (8), 1412–1432. https://doi.org/10.1080/14616688.2024.2305864
Tomassini, L. and Bertella, G. (2023) The human gaze at animals and the missing animal gaze in tourism studies. *Current Issues in Tourism* 27 (3), 350–355.
Tomassini, L. and Cavagnaro, E. (2020) The novel spaces and power-geometries in tourism and hospitality after 2020 will belong to the 'local'. *Tourism Geographies* 22 (3), 713–719.
Tomassini, L. and Lamond, I. (2022) Rethinking the space of tourism, its power-geometries, and spatial justice. *Journal of Sustainable Tourism* 31 (12), 2825–2838.
Urry, J. (1990) *The Tourist Gaze: Leisure and Travel in Contemporary Societies.* Sage.
West, S., Haider, L.J., Stålhammar, S. and Woroniecki, S. (2020) A relational turn for sustainability science? Relational thinking, leverage points and transformations. *Ecosystems and People* 16 (1), 304–325.
West, S., Haider, L.J., Hertz, T., Mancilla Garcia, M. and Moore, M.L. (2024) Relational approaches to sustainability transformations: Walking together in a world of many worlds. *Ecosystems and People* 20 (1), 2370539.
Wheeler, K.M. (1999) *A Discourse of Wonders: Audience and Performance in Ovid's Metamorphoses.* University of Pennsylvania Press.
Yeoman, I. (2012) *2050 – Tomorrow's Tourism.* Channel View Publications.
Yeoman, I. and McMahon-Beattie, U. (eds) (2020a) *The Future Past of Tourism: Historical Perspectives and Future Evolutions.* Channel View Publications.
Yeoman, I. and McMahon-Beattie, U. (2020b) Turning points' in tourism's development: 1946–2095, a perspective article. *Tourism Review* 75 (1), 86–90.
Yeoman, I. and McMahon-Beattie, U. (2023) Future past of tourism: Critical reflection's on the rise of tourism futures. *Tourism Geographies* 27 (3&4), 476–492. https://doi.org/10.1080/14616688.2023.2270937

3 A Posthuman Theorisation of Otherness and *Othering* in Tourism and Migration

Elisa Burrai

Introduction

Capitalist, neoliberal market economies, the rise of new technologies and an increased sociopolitical instability have impacted the resurgence of beliefs about national identities and negative feelings towards different cultures, religions and ethnicities. In such global landscape characterised by determinism through demarcation between 'us' and 'them'/familiar and strange, xenophobia, racism and nationalism have become distinctive features. Hence, in current times, the geopolitical redefinition of territories and strengthening of national borders represents the 'schizoid logic of expansion and contraction, mobility and control, homogenisation and differentiation, which is definitional of the global world order' (Braidotti, 2005: 175).

In this chapter, I focus on the interconnection between migration and tourism, with a focus on the notion of 'other' and on processes of becoming 'others' in relation to the changes and challenges of political and global mobility in a posthuman world. In doing so, this chapter discusses the implications for tourism futures by identifying a utopian and dystopian scenario (Bergman *et al.*, 2010). Adopting a posthuman lens, processes of othering and otherness are examined through the explanation offered by Sara Ahmed (2000) whereby we can have an *Ontology of Strangers* that hinders the possibility of detaching ourselves from the taken-for-granted 'stranger's status as a figure' (2000: 3). Hence, strangers are perceived as such because they are considered in opposition to what is perceived as 'familiar' or 'ordinary'. In other words, public perceptions of strangers are constructed in opposition to the ordinary highlighting how, not only nationalist discourses that associate strangers to unsafety advocate for the expulsion of strangers from hosting territories but also discourses on multiculturalism that advocate for the integration of the stranger welcome [them] 'as the origin of difference' (Ahmed, 2000: 4). Thus, if on the one hand strangers are associated with chaos (Bauman, 2016), on the other hand discourses on welcoming societies reinforce the alterity that characterise ideas and perceptions of 'strangerhood' perpetuating histories of determination which made them excluded in the first place.

Building on Rosi Braidotti's (2019) explanation of posthumanism and Sara Ahmed's (2000) conceptualisation of stranger fetishism, I will provide theoretical insights on how these converge in the field of tourism and how they take shape in possible future tourism scenarios. I do so to advance understanding on the intersection between migration and tourism in order to find possibilities to 'energize projects that express generative narratives and do not wallow in the rhetoric of the crisis. Especially when the crisis in question is to a certain extent the lament of white European cultures feeling vulnerable after anthropogenic global risks are likely to affect them. They need to develop some decolonial perspectives' (Braidotti, 2019: 69).

The aim of this chapter is to examine how and through which processes migration and tourism intersect in a posthuman space. Space here is understood as relationally constructed, it is both sociological and physical, in constant evolution and determined by the (co)existence of a multiplicity of subjects (Massey, 2005). In line with this, I attempt to rethink tourism as a collective space where human/non-human encounters embrace the challenges linked to (forced) migration and respond to them by mobilizing global and local actions. To this end, the discussion seeks to unpack the following research questions: Which spaces are shaped through the intersection of tourism and migration in a posthuman world? How has posthumanism reshaped ways of theorizing and experiencing tourism? How does the concept of *stranger fetishism* fit in posthuman tourism with a focus on the future of tourism? In addressing these questions, I attempt to further engage with a theoretically driven and critical understanding of migration that contributes to widening views on its links and overlaps with/in the tourism landscape providing perspectives on both fields (Burrai *et al.*, 2023). Thus, the chapter follows this structure. First, it provides a discussion on posthumanism and the posthuman. Second, the chapter focuses on the concept of 'the other', its developments and how the idea of *stranger fetishism* fits into posthuman spaces. Finally, posthumanism and *stranger fetishism* are interlinked and discussed within the context of tourism to provide insights into the possible futures of tourism.

Posthumanism and the Posthuman

Recently, public debates and scholarly discussions have focused their attention on what constitutes a 'human' in a posthuman time (Braidotti,

2019). The posthuman condition is understood as the convergence of 'the material and the discursive manifestations of the mutations that are engendered by advanced technological developments (am I a robot), climate change (will I survive), and capitalism (can I afford this)' (2019: 2). Posthumanism enables to redefine the notion of the human in the 21st century. Accordingly, in such a complex scenario, the question of what constitutes the human in this century opens up a plethora of interpretations that are not universal but, instead, culture-specific. Human is not a neutral term. It implies certain degrees of power, inclusion and exclusion which result in perceiving some humans as less human than others. This is the case, for example, of forced migrants such as refugees or asylum seekers. Being human, according to Braidotti (2017), is neither *homo universalis* nor *anthropos*. Instead, humans are embodied and embedded, relational and affective transversal subjects (Braidotti, 2019: 11). Therefore, in posthumanism, people should be rethought in terms of their 'embodied singularity' (Braidotti, 2005: 174), yet, in spite of their heterogeneous nature, the collective subjects are '(all) in this together' (Braidotti, 2019: 54).

A posthuman lens enables to reflect on where we are at as humans and what is happening to us given the complex global transformations and challenges. These include technological advancements (e.g. cyborgs), evolution (e.g. adaptation) and ecology (relation between human/nonhuman). Thus, in the (open) notion of being human, emphasis is placed on the becoming of transversal subjects. Becoming is central to understanding how subjects can deal in an affirmative manner with the crises and challenges of our time. Processes of becoming (what we want to become; what we are capable of becoming) address the fluid notion of identity, which is not static but that can transform and that it is open to a multiplicity of relations and conditions. Through the acknowledgment of the fluidity, transversality and heterogeneity of being human, I question how perceptions of otherness and processes of othering come to the fore within the tourism landscape. In doing so, I embrace a posthuman view that seeks to move away from established boundaries and conventional dichotomies that are part of the human thinking. As such, through a posthuman lens, we can rethink the figure of migrants not as others linked to chaos in opposition to us as (dominant) familiar; instead the focus should be placed on the relations, experiences of and intersections between subjects that are part of global, networked systems characterised by flows of people, goods, technologies, animals and non-human entities. Therefore, within posthuman spaces, relations can be rethought in a way that disrupts uneven spatiality characterised by geometries of power and unjust relations. Instead, the notion of (posthuman) relational space, highlights multiplicity and heterogeneity, 'while prompting affective encounters and actions of affirmative ethics' (Tomassini & Lamond, 2023: 2827). The acknowledgement of processes and the deconstruction of binaries enable to address the 'intersections between them [subjects and categories] in various forms of hybridization' (Cohen, 2019: 418) which, in turn, influences ways of experiencing/doing tourism with/for and by migrants.

Thus, a posthuman approach to understanding the complex link between tourism and migration enables to identify new ways of becoming, existing and newly formed subjectivities that co-exist within the material world we inhabit. Accordingly, despite the sociopolitical, economic, ecological and technological challenges, a posthuman approach to examine migration and tourism provides a navigational tool to explore how collectively we can mobilise alternative practices to resist 'the opportunistic epistemic speed of advanced capitalism' (Braidotti, 2019: 156). This is what Braidotti refers to as 'affirmation' (2019), that is a collective, through the network of multiple and transversal subjectivities, rejection of self-destruction. This is achieved by finding alternative ways of dealing (collectively) with the woes of our time rather than dismissing them. Therefore, an affirmative posthuman reconceptualisation of strangers – identified as migrants in the context of this chapter – is important for two main reasons. First, it opens up possibilities to collectively forge new ways of co-creating spaces of empowerment and participation in tourism which are meaningful for all the people involved and that 'produce social horizons of hope, or sustainable futures' (Braidotti, 2019: 173). Second, a posthuman theorisation of the stranger enables to reframe its conventional (and colonial) understanding as someone that is different from a Eurocentric male perspective detached from 'historical broader asymmetrical power relations that marks some strangers, stranger than others' (Marotta, 2021).

Posthumanism in Tourism

Although there is an increased adoption of posthumanist approaches in tourism, for example through the adoption of Latour's Actor Network Theory (Guia & Jamal, 2020), there is still a paucity of studies that have used a critical posthuman lens to examine modern tourism. More specifically, the use of posthuman theories within the field of tourism should further engage with and challenge the ontological and ethical underpinnings of modern tourism, largely based on dominant anthropocentric and Western capitalist, neoliberal economic systems.

In line with this, recent publications have adopted a posthuman approach to reflect on the nexus between tourism and animals (Cohen, 2018), tourism, justice and ethics (Guia, 2021), ecotourism and animal ethics (Thomsen *et al.*, 2023), and in the field of hospitality (Guia & Jamal). Additionally, a recent special issue (Kumm *et al.*, 2019) on *Posthumanism and Leisure Studies: Inclusions, Expressions, and Criticisms* offers comprehensive insights into the multiple ways posthumanism has been adopted, and its potential within the field of leisure studies. In addition to this, with a specific focus on technology, transhumanism has been discussed in relation to its possibilities to enter the cosmos (Cohen & Spector, 2020). Furthermore, through a posthuman lens applied to the sociology of space, Tomassini and Cavagnaro (2023) examined the local turn in tourism whereby a polarised understanding of hosts and guests/tourists and residents is abandoned. Instead, as Tomassini and Cavagnaro (2023) emphasise, examples of (a posthuman) affirmative ethics in tourism include the existence of local, small-scale projects which take place in Europe and that are co-developed and led by migrants (Ormond & Vietti, 2022; Burrai *et al.*, 2022).

Although these initiatives demonstrate the possibility of meaningful encounters and participation particularly of those actors often excluded or marginalised, there is also the risk that these types of tourism experiences crystalise conventional categories of the 'poor' migrant, emphasising the ontology of the stranger (Ahmed, 2000). As such, the migrant could be regarded as a person who needs to be integrated and welcomed in the host society by the resourceful (familiar) host or, as in the context of refugee-led tours, by privileged tourists. To mitigate this risk, affirmative ethics should be intertwined with the theorisation of strangers as 'others' proposed by Ahmed (2000), whereby we can move away from an ontology of strangers fixed in time and space, acknowledging the many possibilities of belonging, being excluded and of becoming. In order to address this latter point, I refer back to posthumanism as a useful tool to uncover the types of subjects that we are capable of becoming through collective actions and relations within the remits of a world in constant evolution and under challenging pressures.

Posthuman Understandings of the *Ontology of Strangers*

Displacement is a common feature of postmodern societies. It occurs across different temporalities and spaces (i.e. environments and societies). Displacement and migration involve human and non-human mobilities and are the responses of environmental devastation, political instability, oppression, inequality and broken democratic systems (Burrai *et al.*, 2023). In this context, it is important to rethink ways in which human/non-human displacements can be meaningfully addressed in an attempt to forge through affirmative ethics, hopeful, sustainable futures. Tourism in this respect can play an important role because it is a complex industry and field whereby a multitude of (im)mobilities, experiences and identities intersect and transform giving shape to ways of being and new ways of becoming (Braidotti, 2019). Furthermore, tourism has the potential to engage with and influence wider institutional systems (i.e. global and local policies and regulations) opening up possibilities of rapture from neoliberal, capitalist, exploitative and growth-focused approaches to local developments (Braidotti, 2019).

A posthuman lens to displacement with a focus on migration enables to distance current modes of thinking from taken-for-granted binaries. Posthumanism is also thought, for the purposes of this chapter, as a theoretical tool to critically address and engage with social and environmental injustices to pave the way for meaningful changes forged by a collective creative and imaginative vision of just and equitable futures. Collectively there is the potential to creatively and disruptively dismantle Eurocentric, white, male, neoliberal hegemonic ideologies and practices in favour of alternative, ancient indigenous epistemologies and ontologies. Hence, in the tourism field, collective efforts of different subjects should reshape notions of power and knowledge distancing from masculine and Eurocentric ones. Far from being utopian, tourism can contribute to addressing the challenge of racist and violent approaches to displacement and migration through acts of (collective) affirmative alternatives that are empowering and inclusive. Importantly, through collective acts 'we can extract ourselves from this sad state of affairs, work through the multiple layers of our exhaustion, and co-construct different platforms of becoming' (Braidotti, 2019: 181).

Displacement and migration are linked to specific territories, temporalities and power. Hence, different types and modes of (im)mobilities need to be considered in relation to their hybridity and subjectivities (Braidotti & Passerini, 2007). Additionally, through migratory processes, notions of *sameness* and *otherness* come to the fore. In particular, they are constructed within Western-centric sites of power, whereby universal notions of whiteness and masculinity are the defining traits of dominance, and as such universally accepted, sameness. Notions of others and processes of othering are, therefore, developed in opposition to the dominant 'same'. Yet others, othering and sameness are characterised by asymmetries of power and they are linked to histories of colonialism and current trends of populism and hyper-nationalism.

Processes of othering are shaped, according to Ahmed (2000), by proximity and distance in (colonial) encounters. In other words, 'others become strangers (the ones who are distant), and "other cultures" become "strange cultures" (the ones who are distant), only through coming *too close to home*, that is, through the proximity of the encounter or "facing" itself' (Ahmed, 2000: 12). Hence, transnational (im)mobilities of people, goods and capital, shape one's understanding of others based on a present proximity (i.e. too close to home) and on a past distance (i.e. being outside the community). The *Ontology of the Stranger* (as 'other'), or *a priori* differentiation, therefore, exists before the encounter with the other takes place. In constructing narratives based on sociopolitical imaginaries of 'the other', there are different degrees of otherness. As such, some 'others' are more familiar than stranger 'others' (Ahmed, 2000) in relation to multicultural discourses of integration, acceptance and ideas of national identity. It is only through a rapture from fixed and pre-existing forms of differentiation that it is possible to forge new ways of becoming based on alignment rather than a merger (Ahmed, 2000).

Posthuman understandings of migration in tourism to overcome an *Ontology of Strangers*

Public discourses on migration, displacement and categories of migrants are often politically-led and socially constructed. Recent political debates and national policies have focused on the urge of responding to fluxes of migrants and to the so-called refugee crisis. Some of the proposed solutions of national governments included holding migrants in Albania or Rwanda; controlling borders through Brexit in the context of the UK; or using the border wall between Mexico and the US to discourage migration. Considering this challenging scenario, recent publications in the tourism field have looked at ways in which the industry can play a meaningful role to address xenophobic and nationalist approaches to migration (see, e.g. Burrai *et al.*, 2022; Musaro & Moralli, 2019; Ormond & Vietti, 2022).

Tourism and migration are complex and often interrelated phenomena (Burrai *et al.*, 2023) in that they involve different types of mobilities, (uneven) access to resources and a wide array of (im)mobile stakeholders. Conventional perceptions of the two, however, differ in that tourism is usually seen through a positive lens, with tourists benefitting the hosting destination. Instead, migration is often underpinned by negative connotations, with migrants being considered as 'invaders' in the receiving countries; vulnerable; passive and marginalised actors (Musaro & Moralli, 2019). Literature on the links between leisure, hospitality, tourism and migration is extensive. It examines different types of migration and migrants within a variety of geopolitical contexts. Previous theoretical approaches have focused on mobilities of forced migration (Bianchi *et al.*, 2020), mobilities of justice (Sheller, 2021) and critical reconsiderations of shared urban spaces for inclusion and justice (Ormond & Vietti, 2022). Additionally, in examining different types of migrations focus has been placed on labor migration, return migration, entrepreneurial migration, retirement migration and second homes (Williams & Hall, 2000; Truly, 2002). Furthermore, migration has been examined through the consideration of temporary migrants (Back, 2020), seasonal workers (Sun *et al.*, 2020) and (digital) nomads in regard to particular forms of non-permanent migration (Mancinelli & Molz, 2023).

In tourism studies it has been noted how migrants are often associated with precarious work, that is unregulated and insecure (Robinson *et al.*, 2019). Importantly, the intersectionality of migrants which includes their gender, ethnicity, race and class has the potential to strengthen the link between the status of 'being a migrant' and vulnerable/precarious lives. As noted, this is particularly noticeable in the hospitality industry because of its informal and low-skilled nature (Lugosi & Ndiuini, 2022). The nexus between migration and tourism has also been examined from the perspectives of Arab/Muslim women visiting relatives in the West through the application of stigma theory (Moufakkir, 2020). Accordingly, in this study. the 'other' is associated to the tourists (i.e. solo women from Morocco visiting relatives in the Netherlands). As such, tourists carry negative connotations because they represent the stigmatised and stereotyped immigrants (Moufakkir, 2020). In a similar study, AlSaleh and Moufakkir (2019) highlight how less conventional types of mobilities – that from the Global South to the Global North – impact on perceptions and social constructs of the 'other'. In a context of increasing populism and xenophobia towards (some) migrants, stigmatised perceptions of tourists from destinations in the Global South have negative impacts on their tourist experience fostering the desire to end their visit and return back to their countries of origin (AlSaleh & Moufakkir, 2019).

Moreover, literature on tourism and migration has discussed how different (yet well-defined) categories of migrants interface with tourism, as both experience and practice, limiting critical acknowledgements of the multiple possibilities, experiences and perceptions of being migrants within tourism spaces. Hence, as Burrai *et al.* (2023: 530) argue, tourism studies should contribute to advance 'perspectives that can complement and complicate established discussions on tourism and migration which often address their relationship in dichotomous ways (e.g. tourism as site of inequality or resource for migrant integration)'. Similarly, Bloch and Adams (2022) in their edited book, advance fluid thinking on the relation between tourism, migration and refugee studies in an attempt to de-compartmentalise the notions of tourism and migration. In an attempt to problematise seemingly fixed boundaries often attributed to the notions of migration and tourism, Simoni (2022), in his work, reflects on when one is a 'tourist', a 'migrant' and a 'Cuban' contributing understanding on the concepts of belonging and identity in the context of Cuba. Simoni (2022) sheds light on 'when, why and how these categorisations come to matter, how they bring about or hamper certain modes of being and becoming, as well as the possibility of negotiating, resisting and overcoming them' (Simoni, 2022: 46). As such, different types of mobilities and categories of migrants become blurred within the context of tourism where roles, motivations and encounters provide multiple possibilities of being and becoming. This is evident, for example, for members of diasporas that can become tourists or for refugees that can become tourism entrepreneurs (Bloch & Adams, 2022).

In unpacking the multifaceted, embodied and affective possibilities of becoming in the posthuman world, there is the need to move away from categories of 'others', 'otherness' and process of 'othering' as these imply that there is – in contrast – someone in a position of power that represents harmony, stability and familiarity. Accordingly, future tourism scenarios should be forged by the multiple (relational) possibilities generated by the intertwined notions of migration and tourism in a way that goes beyond demarcated patterns.

Currently, the familiar is usually embodied by the notion of sameness created by the West. Tourism, therefore, could be the catalyst to forge positive futures for humans, non-humans and, in broader terms, destinations through the acknowledgement of a multitude of subjectivities which are part of collective projects that prioritise wellbeing, justice and equality. Hence, new ways of thinking and doing tourism should operate on the alignment of coexisting diversity rather than on 'merging' the other in the host society. This implies that we move away from the notion of integration and instead, focus on acknowledging diversity and to work with each other on the basis of it (Ahmed, 2002). This enables to get closer to 'other others' (Ahmed, 2002) making possible a collective politics that 'is what must be worked for, rather than being the foundation of our collective work' (2002: 180). Collectivities, according to Ahmed (2002), 'are formed through the very work that we need to do in order to get closer to others, without simply repeating the appropriation of "them" as labour, or as a sign of difference' (2002: 180).

In tourism, collective politics can take shape in projects whereby migrants, tourists, residents encounter each other (proximity) but that does not allow a merger (e.g. integration). Instead, collective politics recognise the heterogeneity of migrants' subjectivities within localities (e.g. a tourist destination) and in relation to the globe – this latter association, thus, enables to move away from discourses on the refugee crisis, for example, which contribute to create stigmatised and stereotyped images of 'others' that need either to be kept distant through border control and nationalist politics or welcomed (as different others) and integrated in hosting societies (to achieve sameness). Therefore, in order for collective politics to provide a meaningful alternative to Western, colonial and heteronormative processes of othering and traditional forms of social differentiation, it is important to work on subjectivities which will be 'refined in what we have in common' (Ahmed, 2002: 177). Migration in/and tourism should be addressed through working with a multitude of subjectivities in a way that – through encounters – we are able to get closer to those that are other than 'the other' (Ahmed, 2002). These encounters should be rooted in activism and a type of politics based on collective action for affirmative ethics (Braidotti, 2019) to forge hopeful, alternative futures. Tourism futures in regard to the nexus between the industry, its practices and migration should be much more politically positioned in a way that this can create a rapture with universalism and cultural relativism based on a western set of knowledge and bounded by capitalist modernity. Such rapture can open up the possibility of a type of activism 'assumes an intimate relationship between these categorisations and politics (between being and acting). Thinking about how we might work with, and speak to, others, or how we may inhabit the world *with* others, involves imagining a different form of political community, one that moves beyond the opposition between common and uncommon, between friends and strangers, or between sameness and difference' (Ahmed, 2002: 180).

Conclusions: Exploring Utopian and Dystopian Scenarios

This chapter contributed theoretical knowledge to the nexus between tourism, otherness and othering. It drew on the posthuman examination of the concept of otherness and the process of othering in relation to migration in/and tourism and on the notion of '*Stranger Fetishism*' or '*Ontology of Strangers*' discussed by Sara Ahmed (2000: 1–3). Accordingly, 'the stranger is produced, not as that which we fail to recognize, but as that which we have already recognized as a "stranger"' (2000: 3). Indeed, in (modern) Western public discourses, migrants are associated to chaos and instability (Bauman, 2016). As such, fluxes of migrants need to be managed and controlled by national governments as they are seen as threats to the local order, harmony and social cohesion.

Through the use of a posthuman lens and the notion of 'others' elaborated by Ahmed (2000), migration is considered in fluid terms moving away from dichotomous and often static considerations. According to Dahinden (2016) migration is institutionalised as part of modern nation states (2016: 2208). The migration agenda should be de-migranticized as diversity linked to migration is understood in relation to borders and cross-borders movements (Dahinden, 2016). To further understand on Ahmed's view of otherness (2000) as a fluid and holistic phenomenon, I framed migration in tourism adopting Rosi Braidotti's (2019) posthuman view on the meaning of otherness and on the process of othering. This is based on the shift from the Anthropocene to a more holistic, encompassing one whereby human and non-human relations and interactions must be taken into account when reflecting on the role tourism plays in relation to othering and otherness. Indeed, adopting a posthuman lens to understand modern (im)mobilities within the field of tourism enables to gain a 'position that transposes hybridity, nomadism, diasporas and creolization processes into means of re-grounding claims to subjectivity, connections and community among subjects of the human and the non-humankind' (Braidotti, 2019: 50).

A posthuman way of thinking enabled us to engage with the crises of our time, including displacement and migration, not in a dismissive manner through negation of the context but in a hopeful one which opens up possibilities of sustainable and inclusive futures. Hence, a posthuman understanding of individuals as heterogeneous entities which are part of broad human/non-human systems allowed for a reconsideration of diversity hooked to collective subjectivities which – in spite of asymmetrical positions of power, exclusion and inclusion – inhabit the same distraught world. Tourism represents a useful context whereby differences converge and can be acknowledged in a way that enables commonalities to emerge. In a future scenario, tourists, migrants, residents and other tourism stakeholders can – therefore – overcome division based on the (sociopolitical) construct of 'the other' valuing and engineering collective, political and grassroot action that creatively contest the dominant institutional and ideological discourses which often underpin tourism policies and practices.

Drawing on Bergman *et al.* (2010) ontological reflections on future studies and forecasts, the chapter envisages two types of future scenarios, a utopian and a dystopian one, both aiming neither to make truth claims, nor explanatory ones.

The utopian scenario

In a utopian scenario, diversity is (harmoniously) embedded in the subjectivities of individuals that work together to forge new ways of doing tourism. In this context, local communities cooperate to develop a tourism which is well-integrated with other socioeconomic activities. In this utopian scenario, local, hybrid communities adopt, and shape localised, small scale tourisms that are close to the issues and challenges brought to the fore by a time of intersecting ecological, technological, economic and social crises. Tourism can become a vehicle to expose locals and tourists to the multiple possibilities that working together with/for nature can generate and to challenge the idea of bounded spaces (borders) through dispossession, land rights and ownership. These future tourism spaces where relationships and processes unfold, could be characterised by (mobile) networks of people (tourists, migrants, residents) that use a system similar to 'guerrilla gardening' (Reynolds, 2008) as a leisure platform to rethink borderless spaces whereby collectively we look after the environment and learn from it.

The dystopian scenario

In a dystopian scenario, tourism will become a much more undemocratic and uneven industry. In this scenario, there will be the exacerbation of welcoming only certain types of mobilities and accessibility to resources. In line with this, a dystopian technological surveillance could be employed to control movements and protect national borders exacerbating division, fear and inequality. The use of biometric data, drones and geolocation technologies can, therefore, reconfigure flows of uneven mobilities strengthening national constructs around otherness and sameness.

In envisaging sustainable and meaningful tourism futures whereby diversity becomes the catalyst of collective work *with* each other, it is recommended that future empirical studies unpack the link between posthumanism, migration and tourism further. In line with this, it is suggested that further research on the intersection of these topics focuses on specific examples of tourism projects that enable to reconsider migrants not as strangers but as part of a collective posthuman approach to hopeful futures.

References

Ahmed, S. (2000) *Strange Encounters: Embodied Others in Post-Coloniality.* Routledge.

AlSaleh, D. and Moufakkir, O. (2019) An exploratory study of the experience of VFR tourists: A culture shock perspective. *Tourism Geographies* 21 (4), 565–585.

Back, A. (2020) Temporary resident evil? Managing diverse impacts of second-home tourism. *Current Issues in Tourism* 23 (11), 1328–1342.

Bauman, Z. (2016) *Strangers at Our Door.* Polity Press.

Bergman, A., Karlsson J.C. and Axelsson, J. (2010) Truths claims and explanatory claims – an ontological typology of futures studies. *Futures* 42 (8), 857–865.

Bianchi, R.V., Stephenson, M.L. and Hannam, K. (2020) The contradictory politics of the right to travel: Mobilities, borders & tourism. *Mobilities* 15 (2), 290–306.

Bloch, N. and Adams, K.M. (eds) (2022) *Intersections of Tourism, Migration, and Exile.* Routledge.

Braidotti, R. (2005) A critical cartography of feminist post-postmodernism. *Australian Feminist Studies* 20 (47), 169–180.

Braidotti, R. (2017) Posthuman, all too human: The memoirs and aspirations of a posthumanist. *Tanner Lectures, Yale University* 1–2

Braidotti, R. (2019) *Posthuman Knowledge* (Vol. 2). Polity Press.

Braidotti, R. and Passerini, L. (2007) On becoming Europeans. In L. Passerini, D. Lyon, E. Capussotti and L. Laliotou (eds) *Women Migrants from East to West: Gender, Mobility and Belonging in Contemporary Europe*, 23–44.

Burrai, E., Buda, D.M. and Stevenson, E. (2022) Tourism and refugee-crisis intersections: Co-creating tour guide experiences in Leeds, England. *Journal of Sustainable Tourism* 1–18.

Burrai, E., Cherchi, C., De Martini Ugolotti, N. and Sterchele, D. (2023) Being at home and away: Transnational entanglements of tourism and migration in Sardinia, Italy. *Journal of Tourism and Cultural Change* 1–15.

Cohen, E. (2019) Posthumanism and tourism. *Tourism Review* 74 (3), 416–427.

Cohen, E. and Spector, S. (2020) Transhumanism and cosmic travel. *Tourism Recreation Research* 45 (2), 176–184.

Dahinden, J. (2016) A plea for the 'de-migranticization' of research on migration and integration. *Ethnic and Racial Studies* 39 (13), 2207–2225.

Guia, J. (2021) Conceptualizing justice tourism and the promise of posthumanism. *Journal of Sustainable Tourism* 29 (2–3), 503–520.

Guia, J. and Jamal, T. (2020) A (Deleuzian) posthumanist paradigm for tourism research. *Annals of Tourism Research* 84, 102982.

Guia, J. and Jamal, T. (2023) An affective and posthumanist cosmopolitan hospitality. *Annals of Tourism Research* 100, 103569.

Kumm, B.E., Berbary, L.A. and Grimwood, B.S. (2019) For those to come: An introduction to why posthumanism matters. *Leisure Sciences* 41 (5), 341–347.

Lugosi, P. and Ndiuini, A. (2022) Migrant mobility and value creation in hospitality labour. *Annals of Tourism Research* 95, 103429.

Marotta, V. (2021) Meeting again: Reflections on strange encounters 20 years on. *Journal of Intercultural Studies* 42 (1), 1–7.

Mancinelli, F. and Germann Molz, J. (2023) Moving with and against the state: Digital nomads and frictional mobility regimes. *Mobilities* 19 (2), 189–207.

Moufakkir, O. (2020) Experience of Arab/Muslim women visiting relatives in the West and the management of stigma by association. *Tourism Management* 78, 104073.

Massey, D. (2005) *For Space*. Sage.

Musaro, P. and Moralli, M. (2019) De-bordering narratives on tourism and migration. A participatory action-research on two innovative Italian practices. *Italian Journal of Sociology of Education* 11 (2), 147–173.

Ormond, M. and Vietti, F. (2022) Beyond multicultural 'tolerance': Guided tours and guidebooks as transformative tools for civic learning. *Journal of Sustainable Tourism* 30 (2–3), 533–549. https://doi.org/10.1080/09669582.2021.1901908.

Reynolds, R. (2008) *On Guerrilla Gardening: A Handbook for Gardening without Boundaries*. Bloomsbury Publishing.

Robinson, R.N., Martins, A., Solnet, D. and Baum, T. (2019) Sustaining precarity: Critically examining tourism and employment. *Journal of Sustainable Tourism* 27 (7), 1008–1025.

Sheller, M. (2021) Reconstructing tourism in the Caribbean: Connecting pandemic recovery, climate resilience and sustainable tourism through mobility justice. *Journal of Sustainable Tourism* 29 (9), 1436–1449.

Simoni, V. (2022) Migrant, tourist, Cuban: Identification and belonging in return visits to Cuba. In N. Bloch and K.M. Adams (eds) *Intersections of Tourism, Migration, and Exile* (pp. 45–59). Routledge. https://doi.org/10.4324/9781003182689-3

Sun, J., Ling, L. and Huang, Z.J. (2020) Tourism migrant workers: The internal integration from urban to rural destinations. *Annals of Tourism Research* 84, 102972.

Thomsen, B., Thomsen, J., Copeland, K., Coose, S., Arnold, E., Bryan, H. and Chalich, G. (2023) Multispecies livelihoods: A posthumanist approach to wildlife ecotourism that promotes animal ethics. *Journal of Sustainable Tourism* 31 (5), 1195–1213.

Tomassini, L. and Cavagnaro, E. (2023) Circular *oikonomia*, posthumanism and local space to socialise tourism. In F. Higgins-Desbiolles and B. Chew Bigby (eds) *The Local Turn in Tourism: Empowering Communities* (pp. 54–68). Channel View Publications. https://doi.org/10.21832/9781845418809-009.

Tomassini, L. and Lamond, I. (2023) Rethinking the space of tourism, its power-geometries, and spatial justice. *Journal of Sustainable Tourism* 31 (12), 2825–2838.

Truly, D. (2002) International retirement migration and tourism along the Lake Chapala Riviera: Developing a matrix of retirement migration behaviour. *Tourism Geographies* 4 (3), 261–281.

Williams, A.M. and Hall, C.M. (2000) Tourism and migration: New relationships between production and consumption. *Tourism Geographies* 2 (1), 5–27.

4 Animal Ethics and Cultural Relativism: Bridging 'Otherness' to Find Justice for Non-Human Animals in Tourism Futures

Jessica Bell Rizzolo and David A. Fennell

Introduction

Charting a path for ethical wildlife tourism futures often involves being between Scylla and Charybdis. On one side is the danger of neocolonialism, which can apply a one size fits all approach that disregards cultural variation; however, on the other side, extreme cultural relativism can fail to protect the well-being of nonhuman animals used in tourism (von Essen *et al.*, 2020). In this chapter, we work towards a vision of justice for nonhuman animals that bridges otherness. First, we provide a brief overview of captive elephant-based tourism in Asia, and how this form of tourism presents complex cultural and ethical dilemmas. Second, we examine the notion of justice in animal-based tourism and how justice can be applied in cases of cultural variation/across cultural difference. We do this through the application of two frameworks: the scales of justice framework (Fennell & Sheppard, 2021) that integrates various approaches to animal ethics in order to weigh the degree of justice afforded to nonhuman animals across cultures, and the notion of just sanctuaries (developed by Speiran, 2024, following Fennell & Sheppard, 2021). In so doing, we draw upon notions of otherness and space outlined in Jones (2004), Tomassini *et al.* (2022) and Tomassini and Lamond (2023). These include the concepts of a more-than-human 'spatial formations', spatial justice, and the 'ethically visible' other. Finally, we discuss the potential future(s) of elephant tourism and scenarios (based on principles described by Yeoman & McMahon-Beattie, 2023 and Milestad *et al.*, 2014) for this form of tourism. We focus on how elephant tourism futures might be informed by the intersection of the global and the local, resilience and diversification, and on how we can build preferred, sustainable futures for elephants (Dredge, 2022).

Captive Elephant-based Tourism, Otherness and Space

In numerous countries in Southeast Asia, Asian elephants are in a 'strange dichotomy' when it comes to tourism (Szydlowski, 2022: 128). Wild Asian elephants are subject to numerous protections due to their endangered status and are highly prized by tourists. On the other hand, the long-held use of captive-held elephants in the region (Sukumar, 2011) has led to the notion of 'domesticated' elephants (Rizzolo & Bradshaw, 2016, 2018). These captive-held elephants often have fewer legal protections than their legal conspecifics and are used in tourism for entertainment as in the elephant camps of Thailand, venues concentrated in the northern part of the country where elephants are held in captive conditions for riding and/or circus-type shows (Bansiddhi *et al.*, 2020a, 2020b). At other times, captive elephants are used as transportation to view other wildlife, as in Chitwan National Park in Nepal (Pradhan *et al.*, 2018). While captive elephant tourism is a primary economic driver in the region, it is also a source of intense debate due to the diverse working conditions and treatment of elephants in the industry (Bansiddhi *et al.*, 2020a, b).

Tourist preferences and satisfaction can directly lead to improved welfare standards at elephant tourism venues (Flower *et al.*, 2021a). In addition, tourists' attitudes can shift as a result of participation in elephant-based tourism in a manner that benefits elephants (Flower *et al.*, 2021b) and promotes ethical reflection (Taylor *et al.*, 2019). However, tourists are often unable to properly assess animal welfare (Moorhouse *et al.*, 2015) and poor welfare conditions are still rampant. For example, in Thailand, from 2010 to 2020, there was a 70% increase in the number of captive elephants used in tourism and only 5% of these animals were kept under high welfare conditions (World Animal Protection, 2020). Tourists might also transition to different venues (e.g. from elephant riding to elephant bathing) that still have detrimental effects on animal welfare or that are greenwashed to appear more elephant-friendly than they are. Furthermore, the shift among Western consumers towards more elephant-friendly tourism has led some elephant-riding camps to lower prices and welfare standards to attract new tourist markets (Turner & Kusters, 2021), such as Asian tourists who have different views on animal welfare (Moorhouse *et al.*, 2019) and who are more likely to visit elephant tourism venues with lower welfare standards (Flower *et al.*, 2021b).

Therefore, there is a need to ethically examine elephant well-being in these venues, and such well-being is intertwined with the concept of otherness. By 'otherness', we refer to a move towards the 'ethically visible other' described by Jones (2004). Recognizing the other means 'an account of alterity that does not reduce the other to the same... a philosophy of self and other in which both are preserved as independent and self-sufficient, but in some sense in relation to one another' (as cited by Jones, 2004: 272). Nonhuman animals in particular are others who tend to be made invisible, either through being grouped in a collective rather than being seen as individuals, and/or by having the spaces by which they are defined (e.g. 'farm animals' on factory farms) removed from the societal gaze. While Jones (2004) describes this ethical visibility primarily in terms of human-nonhuman relationships, such ethical visibility (or invisibility) can also emerge between different cultural groups, especially when one group 'animalizes' the other one (Roberts, 2003). Therefore, when we refer to the ethically visible other, we refer to this both in terms of bridging cultural variation in attitudes towards captive elephants (e.g. making other ethical frameworks visible) and in terms of centering (making visible) the needs of individual elephants themselves.

The centrality we place on ethics also extends to our definition of 'space' within captive elephant-based tourism. We follow Jones (2004) in his contention that all human-wildlife encounters and the space(s) of those encounters, are ethically charged. As Jones notes, 'within all geographies of human-animal relations (as perhaps in all spatial practices), there are accompanying issues of ethics which are always shadowing those spatialities', so that the space between humans and nonhumans can be framed as 'geoethics' (Jones, 2004: 269). The spatial dimension is evident in the way that the elephants we describe are often termed 'tourism elephants' or 'domesticated elephants', these labels themselves implying that these creatures inherently are part of, or belong to, human touristic spaces. Further, the ethical frameworks we describe can be seen as an effort to delineate the possibilities for different forms of 'spatial justice' for elephants (Tomassini & Lamond, 2023). As noted by Tomassini and Lamond (2023), 'the freedom to move is a stratifying factor of our times' and 'space is imbued with and a product of relations of power' (2023: 2826/2832). The spatial restrictions on elephants used for tourism include physical confinement and limited mobility as well as the psychological detriments of such limits to their agency, such as complex post-traumatic stress disorder (c-PTSD) and impaired social relationships (Rizzolo & Bradshaw, 2016). Such spatial restrictions are not only emblematic of the ways that humans have power over animals, but the spatial enactment of power itself, or the mechanism by which power operates. This spatial dimension is key to the notion(s) of justice and the various scenarios for elephant-based tourism, that we discuss in this chapter.

Captive elephant-based tourism is one example of what Fennell (2023) classifies as NEWT tourism: nature-based tourism, ecotourism and wildlife tourism. Following Fennell (2023) we argue that, given the complex and often overlapping nature of NEWT forms of tourism, much more clarity needs to be placed around these tourism types, especially in an analysis of justice and ethics. Fennell (2023) argues that in the relationship between animals and humans, ecotourism should be based on the first principle that negotiates value around the types of spaces that animals inhabit, as follows:

> Reject as ecotourism all practices that are based on or support animal capture and confinement or other forms of animal use that cause suffering for human pleasure and entertainment. Instead, embrace ecotourism interactions that place the interests of animals over the interests of humans. This would include encounters with free-living animals that would have the liberty to engage or terminate interactions independent of human influence. (Fennell, 2023: 10)

This means that animals held captive in zoos or aquaria should not be viewed as ecotourism, despite the fact that scholars in the past have made this connection (Mason, 2000). In cases where tourists seek to interact with elephants in-situ, this would indicate a perspective of deep justice where the development of these forms of ecotourism has a direct benefit to elephants in the form of habitat preservation through money donated by ecotourism, and where community development is enhanced because of the evolution of higher regard for the health and welfare of elephants. Animal rights would seem to be the most applicable theoretical position here allowing elephants to 'live by their own lights' (Regan, 2004).

New research is casting light on the prevalence of using ecotourism, and wildlife ecotourism, to define and characterize the holding of animals in captive environments like sanctuaries. Given the discussion on the first principle above, Fennell (2024) argues that even though sanctuaries often work towards the welfare and agency of animals, some of which have been badly used in the past, that wildlife ecotourism or even simply ecotourism are not the right terms to describe the animal-human relationship in these spaces. Wildlife ecotourism (D'Cruze *et al.*, 2017; Thomsen, 2022) suggests that another form of ecotourism exists besides the historical usage of the term as, generally, viewing animals in the wild as ecotourism. If we are to use the prefix 'wildlife' before 'ecotourism', then another form of ecotourism exists in the form of captive ecotourism. Fennell (2024) argues that this is inconsistent with the first principle of ecotourism, above.

A better term would thus be sustainable captive wildlife tourism, according to Fennell (2024), with the implication that animals would receive justice in these captive scenarios by being used and commodified in dignified ways that do not harm animals, while at the same time allowing communities and tourism operators to benefit. The question then becomes, within this framework of captive wildlife tourism and its constraints, how exactly justice can operate. To answer this, we turn to the scales of justice framework developed by Fennell and Sheppard (2021), which integrates various approaches to animal ethics in order to weigh the degree of justice afforded to nonhuman animals across cultures.

The Scales of Justice Framework and Captive Wildlife Tourism

The field of animal ethics is founded on several theoretical perspectives that explore the wide range of human-animal interactions, the degree of animal use (or non-use), and the level of care extended to animals in such use (Table 4.1).

Fennell and Sheppard (2021) demonstrated how these different ethical models can inform notions of justice in wildlife tourism, with a focus on elephants. The notion of justice is a complex and often contested one. Rather than considering justice as dichotomous (just or unjust), Fennell and Sheppard (2021) outline a scales of justice framework that is designed to weigh the degree of justice afforded to animals in various uses. This results in different degrees of justice: deep justice, intermediate justice, shallow justice, or no justice, as they apply to the use of elephants in Asia.

Deep justice is similar to an animal rights perspective, which allows animals to live without control or manipulation. Fennell and Sheppard (2021) used the example of elephant viewing in the wild, which is possible in protected areas of Asian countries that have sizable wild elephant populations, such as Udawalawe National Park in Sri Lanka (Travel Connection, 2020). While there is minimal disruption and interference in the lives of the elephants and to their habitat, there is still a commercial element connected with tourists viewing elephants and other wild animals. However, the animals are free to disengage from their interactions with humans, at their instigation.

Intermediate justice can be found in tourism venues where nonhuman animals do not experience overt abuse but where the interests of human visitors still outweigh the needs of the animals. The Millennium Elephant Foundation in Sri Lanka is one example (https://millenniumelephantfoundation.com/). This family-run organization seeks to raise awareness of the plight of elephants and improve welfare (i.e. campaigning against the use of the howdah or rider's throne). Although the organization still permits tourists to ride the elephants, it also encourages walking beside the elephants as an alternative (Rice & Perkins, no date). Intermediate justice is most aligned with the notion of animal welfare: while animal needs are considered and attempted to be met, animals are still used for human ends (e.g. the use itself is not questioned).

Shallow justice for animals would occur when animals are bred for the tourism industry and/or made to perform unnatural behaviors, even if they are provided a captive environment that meets some of their welfare needs (such as food and shelter). Pinnawala Elephant Orphanage (https://us.lakpura.com/pages/pinnawala-elephant-orphanage) in Sri Lanka is one

Table 4.1 Theories of Animal Ethics

Contractarianism	Rawls (1971) argued that most people would argue that although it is wrong to be cruel to animals, there is little scope for expanding the boundaries of the moral circle to include animals because they are not rational and unable to reciprocate acts of goodwill to others. Under such conditions, Rawls argued that we should have 'duties of compassion and humanity in their case...[but]...they are outside the scope of the theory of justice, and it does not seem possible to extend the contract doctrine so as to include them in a natural way' (Rawls, 1971: 512). If we injure or kill an animal, we do not have duties to the animal, which merely has marginal or secondary moral value, but rather to the person who owns these animals in fulfilling the social contract (Mulgan, 2011). We are kind to animals because we want to be, not because we have to be (Garner, 2012).
Ecocentrism	Ecocentrism is a broad-based perspective that places moral value on whole systems rather than the individual entities within these systems. What is morally right, therefore, is the avoidance of harm to the 'integrity, stability and beauty of the biotic community' (Leopold, 1970: 262). Ecocentrism is thus the central guiding principle of land managers, environmentalists and biologists responsible for maintaining a balance in these large systems' living and non-living aspects. As such, placing energy and resources into saving individual animals makes little sense because these resources could be used at the broader ecosystem scale. Furthermore, critics argue that it is both anthropocentric (Rowe, 1997) and informed by utilitarianism, where the greatest good for the greatest number is a priority (Garner, 2004).
Utilitarianism	The teleological or ends-based theory of utilitarianism is concerned with the best consequences for all involved. Singer's (2009) preference for utilitarianism hinges on the concept of equal consideration of interests. The like suffering of one being should be measured against the like suffering of another being. Accordingly, humans and animals prefer to avoid pain and suffering, and the being that experiences the higher level of suffering must be given a higher level of care. Given that animals do, in fact, experience pain and suffering (Niven, 1967), we should factor this into our treatment of animals in order to avoid being speciesist. It is vital to understand utilitarianism because it is aggregative, that is, it does not prioritize the interests of individual animals and, in this sense, is felt to be more akin to animal welfare in weighing costs and benefits for animals and humans (Francione, 1995).
Animal welfare	This broad spectrum of approaches is concerned not with use – because inherent in animal welfare is the notion that animals can be used – but rather with the quantity and quality of care of animals under such use. Models such as the Five Freedoms ensure that hunger and thirst, discomfort, absence of pain, injury and disease, absence of fear and distress, and the ability of the animal to express normal behavior are secured. Animal welfare provides a rational system of animal use that can easily be sanctioned by policy and decision-makers, that is use embedded in a culture of care. However, even though the lives of animals are improved under such a system, an entrenched criticism is that it opens the door to guilt-free exploitation to satisfy human needs (Cochrane, 2009).
Animal rights	Adopting an animal rights perspective means that we accept that animals have inherent value (Regan, 2004), are sentient, and should not be used as someone else's property (Francione, 1995). Regan (2004), for example, argues that animals have consciousness, intentionality and sentience, which render them subjects of a life. Regan (2004) further contends that we ought to be egalitarian in our actions by showing equal respect to animals; we should be non-perfectionist, indicating that one animal species is not better than another or others, and we should not use other species as means to our selfish ends. We have a prima facie duty to protect victims of injustice – human and non-human.

such example. While the elephants live in better conditions than other establishments in Sri Lanka, the 'orphanage' has a breeding program and photo opportunities for the benefit of tourists, and the prime focus is on the profitability of the organization (Brown, no date). While this type of facility can theoretically be congruent with utilitarianism if enough humans (such as community members, employees, mahouts and tourism operators) benefit from the operation, it fails to meet the conditions of animal rights and also tends to violate principles of animal welfare, since elephants tend to experience physical and psychological deprivation and violence in order to meet the needs of humans (Alberts, 2018).

No justice occurs in situations where animals experience extreme abuse, such as tourism venues that rely on negative reinforcement/violence to control animals (which of all the theories mentioned, is only congruent with contractarianism). Animals are kept in conditions that are insufficient in terms of their physiological and psychological needs, with entertainment and commerce being the sole purpose behind their captivity. One example is Namuang Safari Park in Thailand, where elephants and other animals are forced to perform humiliating and degrading tricks for the entertainment of paying tourists (author personal communication with a park visitor, 2023).

Further, Speiran (2024) outlines three different types of sanctuary (just, transitional, and greenwashed) that are tied to the scales of justice framework developed by Fennell and Sheppard (2021). This is important to examine in the case of captive elephant tourism in particular, as trends towards elephant-friendly tourism have led to an increase in venues that bill themselves as sanctuaries. However, these 'sanctuaries' vary greatly in their ethical impacts on humans and elephants.

Just sanctuaries have animals that range freely, are non-consumptive, have a deep commitment to justice, improve animal welfare and conservation, do not allow tourism (or limit tourism to numbers that prioritize animal needs), and are accredited. One example is Boon Lott's Elephant Sanctuary in Thailand or BLES (https://www.blesele.org/). While BLES does allow visitors, the number of visitors allowed at one time is extremely small. Visitors watch elephants grazing and socializing (as they would in the wild) at a safe distance. Further, this type of sanctuary helps elephants regain a sense of agency through (a) the ability to engage in natural behaviors and (b) the freedom to choose how and with whom to feed, socialize, and so on. This benefits the psychological health of elephants who show symptoms of c-PTSD due to prior abuse (Rizzolo & Bradshaw, 2016). Thus, this sanctuary not only prohibits the abuse of elephants but actively seeks to restore justice to elephants previously harmed (Rizzolo & Bradshaw, 2018).

Transitional sanctuaries are said by Speiran (2024) to be shifting from anthropocentrism towards sustainability and justice for animals, giving priority to the interests of animals in the form of husbandry, rehabilitation and environmental education, over simply revenue. These are sanctuaries that come in several different forms (volunteer, self-led, guided) and allow interaction with animals through open, closed or hybrid options. One example is the Mahouts Foundation in Thailand (https://www.mahouts.org/). This foundation is a community-led organization that partners with mahouts and elephant owners. While these mahouts and elephant owners could potentially earn more revenue through a traditional (riding-based) elephant camp, they choose to keep their elephants close to home and let them roam in an adjacent forest, where tourists can observe the elephants' natural behaviors. This benefits the community and environment as well; tourists stay in homestays and contribute tourism dollars to the village, and the forest is preserved for tourism rather than being leveled for agricultural or other purposes. While some training methods (the ankus, a drag chain) are allowed at the mahout's discretion, this venue is transitional in that it works to move, in a culturally sensitive manner, to a model of elephant tourism that prioritizes elephants' needs.

Finally, greenwashed sanctuaries are characterized by being consumptive and anthropocentric, have a shallow commitment to justice, allow tourism, endanger animal welfare and conservation, and may include roadside zoos acting as sanctuaries. Pinnawala Elephant Orphanage in Sri Lanka is one example. As noted in an investigation by the animal welfare charity Born Free, male elephants at this facility are chained during musth (an aggressive period associated with reproductive hormones) to the point of wounds to elephants' legs, they are trained with the threat of pain from an ankus (a shark hook with a spike), and elephants are primarily used as photo props to create tourism revenue (Brown, no date). Greenwashed sanctuaries that promote 'selfies' or close contact with elephants are particularly problematic. These venues both naturalize close proximity to wildlife (e.g. 'hugging' or 'bathing' an elephant) and greenwash its implications. They (falsely) indicate that close contact with wildlife is safe, fun, desirable, and culturally authentic (Rizzolo, 2017, 2021a), which is associated with additional exploitative behaviors, such as the consumption of wildlife-based products or souvenirs (Rizzolo, 2021a).

Differentiating between 'real' and 'fake' sanctuaries is an important part of working towards spatial justice for animals in tourism (Tomassini *et al.*, 2022). The differences between these sanctuaries are inherently spatial, both in terms of the animal's ability to experience mobility and agency, but also in terms of the degree to which the animal is seen primarily as a commodity (e.g. a mere product, or an interchangeable member of a species) rather than an individual (Tomassini *et al.*, 2022). In 'fake' or greenwashed sanctuaries, elephants both have fewer opportunities for physical mobility and also are primarily commodities or props for tourists. Further, this links to the notion of 'an account of alterity that does not reduce the Other to the Same' (Jones, 2004). A core reason why fake sanctuaries violate justice for elephants is because they present elephants as willing participants in anthropocentric pursuits (Bell, 2015), such as posing for photographs, 'drawing', playing human games, and so on. In other words, elephants are reduced to the same; they are made to be 'human-like' rather than respected for their essential otherness.

Conclusion: Three Future Scenarios on Captive Elephant-based Tourism

This chapter has provided a structured ethical approach to wildlife tourism, offering a pathway between cultural relativism and ethical absolutism. Wildlife tourism involves interactions with both human and non-human 'others', requiring a nuanced understanding of their needs and capabilities. Ethical frameworks such as scales of justice and just sanctuaries can help ensure that these 'others' are ethically visible (Jones, 2004). In captive elephant-based tourism, space serves as an embodiment of power between humans and elephants. Given elephants' high intelligence and complex social structures, the ability to exercise agency and engage in natural behaviors is essential for their well-being.

Rather than viewing ethics in wildlife tourism as a rigid dichotomy (ethical/unethical), this chapter has emphasized a spectrum of justice – from deep, intermediate, shallow, to no justice. While deep justice aligns with ethical sanctuaries that prioritize elephants' needs, economic and cultural factors often lead to intermediate or shallow justice models. Additionally, some tourists may lack awareness or willingness to prioritize ethical considerations (Moorhouse *et al.*, 2019). A contextualized approach is necessary to balance the realities of local communities, tourism operators and conservation efforts.

This spectrum-based perspective is particularly relevant to elephant tourism in Southeast Asia, where the distinction between 'wild' and 'captive' is often blurred. For example, the Mahouts Elephant Foundation in Thailand utilizes tourism revenue to transition elephants from captivity to semi-wild environments (Baker & Winkler, 2020). These elephants, while still under human management, regain agency and engage in social and natural behaviors. Similar models in Cambodia further integrate human and elephant interests into ethical tourism (Kopnina & Baker, 2023). Moreover, the COVID-19 pandemic has reshaped ethical wildlife tourism, highlighting the interplay between global and local economies, resilience and diversification. In Thailand, tourism declined sharply, leading to innovative alternatives like virtual safaris (Barker & Rodway-Dyer, 2023; Fennell, 2020). While virtual tourism may enhance elephant welfare by reducing direct human contact, it also risks 'undertourism' – where economic dependence on tourism leads to severe financial impacts (Fennell, 2020; Kungwansupaphan, 2021). In places like Surin, Thailand, the crisis exposed economic vulnerabilities tied to international tourism, underscoring the need for diversified revenue streams. Given this shifting landscape, we present three possible futures for captive elephant-based tourism:

Scenario 1: Deep justice, or a utopia for elephants

First, we offer a scenario that uses backcasting or working backwards from a target (Milestad *et al.*, 2014) to imagine a future of deep justice for elephants. This is congruent with how scenarios can be a tool to identify the building blocks for preferred, sustainable futures (Dredge, 2022). In a scenario of deep justice, all elephants used in tourism would have space, agency, and be valued as individuals. Elephants would either be viewed in the wild or, if this is not possible, be in just sanctuaries. In order to achieve this, fake or greenwashed sanctuaries would need to be eliminated through both regulations, or limitations on the ability of these 'sanctuaries' to operate and/or advertise on platforms such as Tripadvisor, and through tourist education, or informing tourists that close contact with captive animals is a sign that elephants' needs are not being prioritized. Since just sanctuaries require more space, it would be ideal for the government and/or tourism agencies to invest resources (land, funds, etc.) in these venues instead of using tourism funds to promote 'elephant camps'. Since these venues accommodate a smaller number of in-person visitors at any one time (compared to poor-welfare elephant camps), visitors would also have the option to go on 'virtual safaris' without the expense of travel.

Scenario 2: Intermediate justice, or a step forward

In a scenario of intermediate justice, while elephants would still be viewed as commodities, there would be increased awareness of the need to reconcile the commercial and ethical aspects of captive elephant-based tourism. As discussed by Tomassini *et al.* (2022), nongovernmental organizations such as World Animal Protection can advance the trajectory of intermediate justice through documenting the conditions of animals used in tourism and advocating for improvements through outreach to governments, tourists and other stakeholders. Intermediate justice would consist primarily of transitional sanctuaries, or sanctuaries that give priority to the interests of animals in the form of husbandry, rehabilitation and environmental education, over simply revenue. While these sanctuaries may lack the full conditions of spatial justice present in deep justice venues, they are venues that are beginning to question and move away from anthropocentric values.

Scenario 3: Shallow justice, or maintenance of the status quo

The final scenario would involve little to no change in the status of captive elephants in tourism. This is an elephant tourism industry in which most venues do not meet even the most basic welfare needs for elephants, and in which mobility and agency for elephants are virtually nonexistent. As documented recently by World Animal Protection (2020), 63% of the elephants examined were in severely inadequate conditions, with only 7% kept in high-welfare, observation-only venues. Further, the World Animal Protection found numerous instances of greenwashed sanctuaries or facilities that use labels such as 'ethical' and 'rescue center' despite their poor welfare conditions.

The model we have outlined can help stakeholders identify situations in which there is the willingness and resources to advance transitional or deep justice for elephants. Acknowledging the interdependence of international and local tourism contexts, and viewing justice as a spectrum, can help make both other cultures and other species ethically visible, and can advance a tourism industry that is sustainable for both elephants and human communities.

References

Alberts, E.C. (2018, July 20) This 'sanctuary' is actually a nightmare for elephants. *The Dodo*. See https://www.thedodo.com/in-the-wild/pinnawala-elephant-orphanage-sri-lanka (accessed April 19, 2024).

Baker, L. and Winkler, R. (2020) Asian elephant rescue, rehabilitation and rewilding. *Animal Sentience* 28 (1), 1–20.

Barker, J. and Rodway-Dyer, S. (2023) The elephant in the Zoom: The role of virtual safaris during the COVID-19 pandemic for conservation resilience. *Current Issues in Tourism* 26 (13), 2221–2234.

Brown, V. (no date) Why we don't support Pinnawala Elephant Orphanage. *Responsible Travel*. See https://www.responsiblevacation.com/vacations/elephant-conservation/travel-guide/why-we-dont-support-pinnawala-elephant-orphanage (accessed April 19, 2024).

Bansiddhi, P., Brown, J.L., Thitaram, C., Punyapornwithaya, V. and Nganvongpanit, K. (2020) Elephant tourism in Thailand: A review of animal welfare practices and needs. *Journal of Applied Animal Welfare Science* 23 (2), 164–177.

Bansiddhi, P., Brown, J.L. and Thitaram, C. (2020) Welfare assessment and activities of captive elephants in Thailand. *Animals* 10 (6), 919.

Bell, J. (2015) There is no wild: Conservation and circus discourse. *Society and Animals* 23 (5), 462–483.

Carroll, A.B. (2004) Managing ethically with global stakeholders: A present and future challenge. *Academy of Management Executive* 18 (2), 114–120.

Cochrane, A. (2009) Ownership and justice for animals. *Utilitas* 21 (4), 424–442.

D'Cruze, N., Machado, F., Matthews, N., Balaskas, M., Carder, G., Richardson, V. and Vieto, R. (2017) A review of wildlife ecotourism in Manaus, Brazil. *Nature Conservation* 22, 1–16.

de Bakker, F.G.A., Rasche, A. and Ponte, S. (2019) Multi-stakeholder initiatives on sustainability: A cross-disciplinary review and research agenda for business ethics. *Business Ethics Quarterly* 29 (3), 343–383.

Donaldson, T. and Dunfee, T.W. (1994) Toward a unified conception of business ethics: Integrative social contracts theory. *Academy of Management Review* 19 (2), 252–284.

Donaldson, T. and Dunfee, T.W. (1999) When ethics travel: The promise and peril of global business ethics. *California Management Review* 41 (4), 45–63.

Douglas, M. (2000) Integrative social contracts theory: Hype over hypernorms. *Journal of Business Ethics* 26, 101–110.

Dredge, D. (2022) Regenerative tourism: Transforming mindsets, systems and practices. *Journal of Tourism Futures* 8 (3), 269–281.

Fennell, D.A. (2013) Contesting the zoo as a setting for ecotourism, and the design of a first principle. *Journal of Ecotourism* 12 (1), 1–14.
Fennell, D.A. (2018) The future of ethics in tourism. In C. Cooper and E. Fayos Sola (eds) *The Future of Tourism* (pp. 155–178). Springer.
Fennell, D.A. (2020) Technology and the sustainable tourist in the new age of disruption. *Journal of Sustainable Tourism* 29 (5), 767–773. https://doi.org/10.1080/09669582.2020.1769639.
Fennell, D.A. (2022a) Bridging the global-local animal-based tourism divide. *Annals of Tourism Research* 96 (C) 96, 103456. https://doi.org/10.1016/j.annals.2022.103459.
Fennell, D.A. (2022b) An animal welfare literacy framework for tourism. *Annals of Tourism Research* 96, 103461.
Fennell, D.A. (2023) Animal health warning labels in nature-based, ecotourism & wildlife tourism. *Journal of Ecotourism* 22 (3), 451–458.
Fennell, D.A. (2024) *Tourism and Animal Ethics* (2nd edn). Routledge.
Fennell, D.A. and Sheppard, V. (2021) Tourism, animals and the scales of justice. *Journal of Sustainable Tourism* 29 (2–3), 314–335.
Fennell, D.A. and Thomsen, B. (2021) Tourism & animal suffering: Mapping the future. *Annals of Tourism Research* 91, 103317.
Flower, E.K., Burns, G.L. and Jones, D.N. (2021a) How tourist preference and satisfaction can contribute to improved welfare standards at elephant tourism venues in Thailand. *Animals* 11 (4), 1094.
Flower, E.K., Burns, G.L., Jones, D.N. and McBroom, J. (2021b) Does the experience make a difference? Comparing tourist attitudes pre-and post-visit towards the elephant tourism industry. *Annals of Tourism Research Empirical Insights* 2 (2), 100025.
Francione, G.L. (1995) *Animals, Property and the Law*. Temple University Press.
Garner, R. (2012) Rawls, animals and justice: New literature, same response. *Res Publica* 18, 159–172.
Jones, O. (2004) (Un)ethical geographies of human-non-human relations. In C. Philo and C. Wilbert (eds) *Animal Spaces, Beastly Places* (pp. 267–289). Routledge.
Kopnina, H. and Baker, L. (2023) Conservation, animal wellbeing, and indigenous participation at an elephant sanctuary in Mondulkiri, Cambodia. *Society & Animals* 1 (aop), 1–22.
Kungwansupaphan, C. (2021) The socio-economic impact of COVID-19 on Khunchaitong Elephant community-based tourism in Surin province, Thailand. *Journal of Mekong Societies* 17 (2), 28–49.
Leopold, A. (1970) *A Sand County Almanac with Essays on Conservation from Round River*. Ballantine Books.
Lindgren, N. and Öhman, J. (2019) A posthuman approach to human-animal relationships: Advocating critical pluralism. *Environmental Education Research* 25 (8), 1200–1215.
Lovelock, B. and Lovelock, K.M. (2013) *The Ethics of Tourism: Critical and Applied Perspectives*. Routledge.
Malloy, D.C. and Fennell, D.A. (1998) Ecotourism and ethics: Moral development and organizational cultures. *Journal of Travel Research* 36 (spring), 47–56.
Milestad, R., Svenfelt, Å. and Dreborg, K.H. (2014) Developing integrated explorative and normative scenarios: The case of future land use in a climate-neutral Sweden. *Futures* 60, 59–71.
Moorhouse, T.P., Dahlsjö, C.A., Baker, S.E., D'Cruze, N.C. and Macdonald, D.W. (2015) The customer isn't always right—conservation and animal welfare implications of the increasing demand for wildlife tourism. *PloS One* 10 (10), e0138939.
Moorhouse, T.P., D'Cruze, N.C. and Macdonald, D.W. (2019) Are Chinese nationals' attitudes to wildlife tourist attractions different from those of other nationalities? *Journal of Sustainable Tourism* 27 (1), 12–33.

Mulgan, T. (2011) *Ethics for a Broken World: Imagining Philosophy After Catastrophe.* McGill-Queen's University Press.
Niven, C.D. (1967) *History of the Humane Movement.* Johnson.
Pradhan, N.M., Williams, A.C. and Dhakal, M. (2011) Current status of Asian elephants in Nepal. *Gajah* 35, 87–92.
Rawls, J. (1971) *A Theory of Justice.* Oxford University Press.
Regan, T. (2004) *The Case for Animal Rights.* University of California Press.
Rice, A. and Perkins, R. (no date) Elephant sanctuaries which we do and don't support. *Responsible Travel.* See https://www.responsiblevacation.com/vacations/elephant-conservation/travel-guide/elephant-sanctuaries-which-we-do-and-dont-support (accessed April 19, 2024).
Rizzolo, J.B. (2017) Exploring the sociology of wildlife tourism, global risks, and crime. In M. Gore (ed.) *Conservation Criminology: The Nexus of Crime, Risk and Natural Resources* (pp. 133–156). Wiley-Blackwell.
Rizzolo, J.B. (2021a) The rise of selfie safaris and the future(s) of wildlife tourism. In G. Bertella (ed.) *Wildlife Tourism Futures: Encounters with Wild, Captive and Artificial Animals* (pp. 57–70). Channel View Publications.
Rizzolo, J.B. (2021b) Wildlife tourism and consumption. *Journal of Sustainable Tourism* 31 (5), 1181–1194.
Rizzolo, J.B. and Bradshaw, G.A. (2016) Prevalence and patterns of complex PTSD in Asian elephants (*Elephas maximus*). In A. Manatunga (ed.) *Asian Elephants in Culture and Nature* (pp. 291–297). Centre for Asian Studies, University of Kelaniya.
Rizzolo, J.B. and Bradshaw, G.A. (2018) Human leisure/elephant breakdown: Impacts of tourism on Asian elephants. In N. Carr and J. Young (eds) *Wild Animals and Leisure: Rights and Wellbeing* (pp. 113–131). Routledge.
Rizzolo, J.B. and Bradshaw, G.A. (2019) Nonhuman animal nations: Transforming conservation into wildlife self-determination. *Society & Animals* 29 (4), 393–413.
Roberts, M.S. (2008) *The Mark of the Beast: Animality and Human Oppression.* Purdue University Press.
Rowe, J.S. (1997) From reductionism to holism in ecology and deep ecology. *The Ecologist* 27 (4), 147–151.
Singer, P. (2009) *Animal Liberation* (Reprint). Harper Perennial.
Soule, E. (2002) Managerial moral strategies? In search of a few good principles. *Academy of Management Review* 27, 114–124.
Speiran, S. (2024) Primates in Proximity: The Lives of Monkeys in Costa Rican Sanctuary tourism. PhD dissertation, Queen's University, Kingston, Ontario, Canada.
Sukumar, R. (2011) *The Story of Asia's Elephants.* The Marg Foundation.
Szydlowski, M. (2022) Elephants in Nepal: Correlating disease, tourism, and welfare. *Journal of Applied Animal Welfare Science* 25 (2), 126–138.
Taylor, M., Hurst, C.E., Stinson, M.J. and Grimwood, B.S. (2020) Becoming care-full: Contextualizing moral development among captive elephant volunteer tourists to Thailand. *Journal of Ecotourism* 19 (2), 113–131.
Tomassini, L., Bertella, G., Grasso, C. and Lenzi, C. (2022) The space of animal justice in wildlife sanctuaries: A posthuman perspective. *Journal of Ecotourism* 23 (1), 38–55. https://doi.org/10.1080/14724049.2022.2122480
Tomassini, L. and Lamond, I. (2023) Rethinking the space of tourism, its power-geometries, and spatial justice. *Journal of Sustainable Tourism* 31 (12), 2825–2838.
Thomsen, B. (2022) Wolf ecotourism: A posthumanist approach to wildlife ecotourism. In D.A. Fennell (ed.) *Routledge Handbook of Ecotourism* (pp. 17–131). Routledge.
Travel Connection (2020, January 20) Top 6 places to see wild elephants in Sri Lanka. *Travel Lanka Connection.* See https://www.travellankaconnection.com/info/top-places-to-see-wild-elephants-in-sri-lanka/ (accessed April 19, 2024).

Turner, D. and Kusters, N. (2021) The paradox of elephant-friendly tourism. In E. Laws, N. Scott, X. Font and J. Koldowski (eds) *The Elephant Tourism Business* (pp. 259–268). CABI.

Vincent, R.J. (1986) *Human Rights and International Relations*. Cambridge University Press.

von Essen, E., Lindsjö, J. and Berg, C. (2020) Instagranimal: Animal welfare and animal ethics challenges of animal-based tourism. *Animals* 10 (10), 1830.

World Animal Protection (2020) *Elephants not Commodities: Taken for a Ride 2*. See https://www.worldanimalprotection.org.uk/cdn/ff/f_m_WnfaSOkqb8SFAgyiWz2F3JZ3wxhGoZzmpmSM16A/1597220651/public/media/Taken-For-a-Ride-2.pdf (accessed April, 19, 2024).

Windsor, D. (2018) Dynamics for integrative social contracts theory: Norm evolution and individual mobility. *Journal of Business Ethics* 149, 83–96.

Yeoman, I. and McMahon-Beattie, U. (2023) Future past of tourism: Critical reflection's on the rise of tourism futures. *Tourism Geographies* 27 (3&4), 476–492.

5 Sugar Shacking Tourism Assemblage: Deterritorialization and *Othering* of Maple Trees in Complex Possible Futures

Dominic Lapointe, Arianne Simard-Picard, Julien Menu, Guillaume Ethier and Laurence Laroque Labossière

Sounds of Spring, Tasting the Hope of Summer

While many signs indicate that spring is near, for maple sugarer's families still working with the traditional steel bucket for harvesting maple water, nothing is indicating it more than the sounds of the maple water dripping out of the spout into the bucket with its recognizable 'plink' (Kimmerer, 2013). Mirroring this cultural awakening after a long winter, the biological process that allows the sugar maple (*Acer Saccharum*) to produce the maple water used to produce maple syrup, toffee and sugar could be described as follows:

> During the periods, when temperatures rise above freezing, positive pressure develops in the tree. This pressure causes the sap to flow out of the tree through a wound (tap hole). During cooler periods, when temperatures fall below freezing, negative pressure (suction) develops, drawing water into the tree through the roots. This replenishes the sap in the tree, allowing it to flow again during the next warm period. Although sap generally flows during the day when temperatures are warm, it has been known to flow at night if temperatures remain above freezing. (Extension, 2019, n.p.)

Tapping maple trees to collect maple water is an age-old practice that is geographically associated with the north-east part of America where the climate conditions allow for the alternation of thawing during the day and freezing at night. It also needs the combination of the presence of *Acer saccharum* or, less frequently used, the *Acer rubra* (Red Maple) and of an 'extended period' of spring weather that alternates over zero Celsius temperatures during the day and below zero temperatures at night. Although the red maple can grow all the way to Alabama, the southernmost areas of commercial maple syrup production are the mountains of Virginia, while the Province of Québec, in Canada, produces around 70% of the global supply of maple syrup of the world.

Maple syrup is associated with springs and a variety of cultural heritage practices. In fact, as straightforward as collecting maple water from trees may sound at first, different areas of production assemble distinctive cultural heritage practices around it.

In the St-Lawrence Valley, where our research takes place, the cultural heritage associated with the production and consumption of maple syrup and sugar is referred to as *Les traditions du temps des sucre,* which literally means the sugar time tradition. Those traditions were officially recognized as immaterial heritage by the Québec government in 2023. This interaction between trees, the *Acer saccharum* and the *Acer rubra*, know-how and immaterial heritage creates a sense of uniqueness, or what could be called a niche product for the Québec tourism industry. In this case, the niche is quite adequate as the concept of niche originated in the field of ecology before migrating to marketing where a niche product describes a highly specialized product (Novelli *et al.*, 2022). Not unlike the ecological niche that emerged at this very specific intersection of resources and conditions in the southern regions of Québec, maple products and the different practices associated with it indeed constitute a niche product that nonetheless plays an integral part in the province's cultural identity.

The tourism possibilities created by the *Acer saccharum* are at the intersection of many phenomena, materials, cultural, climatic, ecological and economic logics, and so forth. We will consider those intersections as an assemblage (Deleuze & Guattari, 1980). Mostly presented through the simple logic of the managerial touristification of a 'niche' activity, such a tourism phenomenon hides a much more complex entanglement than it might look like at first (Lapointe & Muldoon, 2024). Indeed, it entangled humans and non-humans, humans and trees, through spatial and social relations, within an othering process. As such, drawing on ontological reflections on future studies (Bergman *et al.*, 2010) and poststructuralist approaches in critical future research (Inayatullah, 1998), this chapter offers novel perspectives into the challenges of contemporary tourism and its possible futures.

While the question of the 'Other' is central to the tourism phenomenon (Höckert, 2018; Lapointe, 2024; Picard & Di Giovine, 2014), it remains mostly circumscribed to the encounter between the Self and Other in the mobility from one's place to the place of other (Équipe, 2002). While this dimension of the 'Other' is important, the othering process, how others, human and more-than-human, are constituted as different that the dominant subject is seldom addressed. In this regard, we turn to Braidotti (2011) who stresses the role of the othering process as a core of the contemporary capitalism, in which the constitution of the Other is a form of differentiation that leads to the possibility of commodification. The constitution of the Other for commodification goes through a deterritorialization towards relations of commodification.

This chapter aims to expose the complexity of the movement of deterritorialization/reterritorialization that feeds the othering process of the maple tree, more-than-human subject central to the tourism around the sugar time traditions in Québec, Canada, and how tourism creates othering through practices and reterritorialization. To address this complexity, we will first turn ourselves to the assemblage theory. Then, we will expose historical elements of the assemblage, a geographical overview of the tourism of sugar shack and maple trees in the St-Lawrence Valley, and try to delineate what happens in the sugar shack industry in Québec tourism. We conclude with an exploration of the assemblage of tourism and maple syrup in Québec, followed by a critical reflection on its implications for future tourism development. Drawing on utopian and dystopian forecasts in future studies (Bergman *et al.*, 2010) and poststructuralist approaches in critical future research (Inayatullah, 1998), the chapter interweaves Assemblage Theory (Deleuze & Guattari, 1980) with Inayatullah's (1998) Causal Layered Analysis (CLA). By building on their shared poststructuralist foundations, this integration helps theorize and shape possible futures by examining multiple layers of reality and their assemblage.

Assemblage Theory Becoming Method

Let us first turn to Deleuze and Guattari (1980). In their book *Milles Plateaux*, philosopher Gilles Deleuze and psychoanalyst Félix Guattari outline the theory of *agencements* to address the question of multiplicity and futures. This complex, multifaceted theory is sometimes summarized in terms of three axes: territorialization/deterritorialization, coding/decoding and expressive/material. Assemblage allows us to grasp the complexity of heterogeneous relationships that enable a whole to be more than the sum of its parts without being contingent to this sum of parts. The concept of assemblage emphasizes relations of exteriority, where constituent elements cannot be reduced to their function within the whole and can, at the same time, be part of a multiplicity of adjacent assemblages. Territorialization thus refers to the stability of assemblages and deterritorialization to the instability of assemblages. Territorialization stabilizes the arrangement of assemblages, reproducing predictable relationships, while deterritorialization destabilizes the arrangement of assemblages, creating new relationships.

For Deleuze and Guattari, this movement also refers to the notion of Earth: the free surface of the Earth is the plane of immanence, the whole, while territory is a section, a corner and a delimited space. These two notions come together, territorialize and deterritorialize, from territory to Earth and Earth to territory, folding and unfolding with movements, practices, continuities and ruptures. This movement is twofold, with a transcendental and an immanent movement, where transcendental ideals and ideas can be deterritorialized, like the principles of free movement that open borders. Deterritorialization can also be immanent, liberating the individual, the ego, from the Earth, which is then set in motion, becoming the other, reterritorializing itself with new practices. The deterritorializing movement is made through lines of flights, which Deleuze and Guattari qualified as a line of intensity holding virtual potential, possible new assemblages and reterritorializing possibilities.

This reterritorializing movement, which sets new limits for the whole, functions thanks to the internal properties of deterritorialized movements. Like deterritorialized movement, it retains its components and stability, and reorganizes its layout according to its internal qualities, a reorganization that will redefine boundaries and reproducible movements by re-anchoring itself to the Earth, drawing a new territory, a combination of physical and symbolic elements that stabilize together. These transformations in movement send out lines of flights on the plane of immanence, sets of virtualities with a potential future. The movement of territorialization/deterritorialization acts through coding and decoding, in the sense of disentangling the previous code, of the material and expressive matters at play in the assemblage. It is those elements that keep the assemblage moving in constant becoming.

Another important principle of assemblage theory is its intrinsic refusal of reductionism (Briassoulis, 2017; Dovey, 2009). In the context of tourism, it offers a way to expose the complexity of tourism entanglements with the other parts of the place and people where it happens (Lapointe, 2021). Indeed, many critics of tourism research are formulated on assumptions of a reduction of its reach and complexity to resume it to a managerial process (Ateljevic *et al.*, 2007), a location-based analysis (Franklin, 2012), or a disembodied disembedded apolitical phenomenon (Ek & Tesfahuney, 2016). Turning to assemblage theory is one way to take into account those caveats of tourism research and also address tourism as an embedded and embodied phenomenon (Braidotti, 2019) intersecting with many spheres of life (Lapointe & Coulter, 2020).

The method we used to construct this chapter was consistent with the assemblage theory as we built different sets of expressive material, deterritorializing archival knowledge and grounded knowledge to reterritorialize it along its expressive quality within an academic code while keeping its inherent qualities as complex dynamic entanglements. Different sets of knowledge were set in movement. One set was built through literature and archival work. For instance, the second author searched in a body of archival documents for mentions of social, cultural, culinary and tourism activities along the maple sugar-making traditions. The third author delved into the geography of maple and maple sugar making. To identify tourism sugar houses, we search the directories of the different destination management organizations of Québec. We extracted all the sugar houses mentioned in the main tourism DMO of Tourism Québec, and all those mentioned in the online directory of each of the 22 regional DMOs of Québec. We then situated them on a map of the province. From the websites of the listed sugar houses, we identified the different activities they were offering and reported them on an excel sheet. From that sheet we could identify three different codes of activities that they offered. The fifth author conducted two interviews with the managers of a cooperative sugarhouse, interviews that were transcript and summarized in an analytical table along the thematic of the chapter. These heterogeneous materials were brought together in a series of seminars on the subject. These moments offered a chance for each of the authors to present its findings and build on a collective assemblage. This process about building a common understanding of the meaning of the material and the work it operates in the assemblage.

This chapter uses assemblage theory but is also an assemblage in itself. But before exposing the assemblage of sugar house tourism in Québec, we will describe some of the elements that were brought together to build a common understanding of the scholarly assemblage in this chapter.

From the Trees to the Food: Recoding in Time of Sugar-Making Activities

Several ethnological studies mention that the sugaring period was the subject of family gatherings among the First Nations of the semi-nomadic nations of Ojibwa, Chippewa, Anishinaabe, and Innu (Chamberlain, 1891a, 1891b; Friedl, 1956; Henshaw, 1890). Held in the Spring during the sugar moon period, these gatherings had a religious character and were a space and time of exchange when the French colonizers came to barter (Franzen *et al.*, 2018). The sweet properties of the maple were not just a material practice, as it was also linked to the First Nations cosmogony, as Kimmerer (2013) is underlining in the recollection of the myths of Nanabozho. This moment of transition between winter and spring was also a gathering moment when groups would get together and exchange goods, the maple sugar also being a trading good.

Before colonial times, indigenous nations would concentrate the sugar using freezing temperatures and boiling:

> (...) they collected sap in birch bark pails and poured it into log troughs hollowed from basswood trees. The large surface area and shallow depth

> of the troughs were ideal for ice formation. Every morning, ice was removed, leaving a more concentrated sugar solution behind. The concentrated solution could then be boiled to sugar with far less energy required. The freezing nights did the work of many cords of firewood, a reminder of elegant connections: maple sap runs at the one time of year when this method is possible. (Kimmerer, 2013: 67)

In this entanglement with the maple trees, the relationship of indigenous people was through cosmogony and kinship (Kimmerer, 2013). As Franzen *et al.* (2018) stress from archeological evidence in the North Michigan Peninsula, the maple sugar was a trading commodity that Chippewa women would exchange with the Hudson Bay company. The production of maple sugar was done outside in large pots and pans over open fire. This form of production was generalized until the popularization of the evaporator, which was an adaptation of the cane sugar making from the south. First a technical advance, the evaporator became central to the iconography and imaginary of the sugar maple production, especially the wood-fire evaporator (Lange, 2017). The evaporator moved the production inside, in what is generally called the sugar shack: a wooden plank shack with an opening in the roof to let the steam go from the evaporator. Dupont (1975) also makes a difference in the late 18th and early 19th centuries between farmer sugarer and forest sugarer. The latter went on a month-long expedition to temporary sugar camps on lands not belonging to him, while the other went for a week to permanent camps which he owned.

The sugar maker of the forest makes the sugars on crown lands which do not belong to anyone (Dupont, 2004). Others obtain a government permit to extract sugars on a given land. According to Dupont (2004), these are the conditions of the forest camp, which is only a temporary shelter – a hut or a rudimentary shelter – where one boiled maple sap in a cauldron over a wood fire. This description, beyond recollecting the ways of the past, is an example of settler colonial representation, as it erased the indigenous presence and their importance in the maple sugar production by describing crown lands as not belonging to anyone. Reinforcing *terra nullius* allows for settler society to anchor in the land, and in their own terms, the practices learned from indigenous people.

In Québec, the sugar shack is also linked to food traditions. Some state that the sugar shack meal, not to say feast, is a celebration of the end of winter (Gaudreau, 2020). On its side, the ethnogeographer Jean-Claude Dupont (1975, 2004) mentioned that these meals are also linked to religious practices for the French-speaking settlers, as usually sugar seasons are timed with the celebration of carnivalesque Mi-Carême. According to the documents of this ethnographer, the residents of the Beauce region, where most of the maple product is produced, used this moment to break their fast and had some weight to their frame. According to the recollection of the 19th and early 20th century, those moments to get together were mostly limited to the close family (Chapais, 1883; Dupont, 2004; Sulte, 1918).

In 1942, an article in *La Revue des Fermières*, an official publication by the *Cercles des Fermières*[1] published by the Québec Ministry of Agriculture, is recounting how the events are happening. The anonymous text mentions that members of the family from all ages get together to eat the maple products. The text also mentions that this festive meal is on an invitation only base, adding that city dwellers must know people in the countryside to expect an invitation to those sugar parties.

The first mention of the sugar shack as a potential tourism asset comes in a 1952 Provincial government publication titled '*L'érable à sucre du Québec*' (Méthot & Rompré, 1952). The publication invites the owners of sugar maple groves to build what they called a 'Villa Sucrière', which could be translated as a sugar house.[2] According to this document, those sugarhouses: 'could welcome groups of young and older people who desire to taste in situ the tasty maple dishes while enjoying the beauty of the maple grove in the spring'[3] (Méthot et Rompré, 1952: 15). Calling for a development of this type of attraction, the authors argue that it would diversify the revenue stream of sugarer by offering traditional Canadian food and leisure opportunities in the tradition of good French-Canadian hospitality.

Along the Roads where the Tourists Flow

The production of maple syrup is closely associated with the ecosystem and climate conditions appropriate for sugar maple and red maple to grow. The leisure and tourism will share those particular practices with others, led to a deterritorializing movement in the 1950s. As we mentioned before, the Québec government published a guide calling for the building of tourism sugar houses. The publication that suggests how to build sugar houses and what food to offer also suggests building them along the main roads used by tourists, mostly American motorists, who visit the province. This indictment would start to move apart the production of maple products and its consumption. This movement is also seen in regard to the distribution of the *Acer Saccharum*. In Lange (2017), an ethnography of maple production in Vermont, one of the informants mentions some sugarer rebuilding the sugar shack closer to the road to be more accessible to tourists' mobility despite the traditional logic of having the sugar shack at the lowest point of the maple grove to use gravity in the moving of the maple water towards the evaporator.

This movement of deterritorialisation of the tourism sugar shack reterritorializes it along tourism mobility, growing apart from the groves that produce the sugar. In Québec, this movement also de-ruralizes the tourism and leisure dimensions of it. Indeed, by the superimposing of the actual maple productive areas (Map 1) with the sugar house listed by tourism authorities, we could visualize a double movement, with the production moving north and east while the tourism maple experience crystallising in the green belt of the Montréal Metropolitan area and, to a lesser point, the Québec City metropolitan area (Map 2). These patterns of implementation create a 'donut effect' around the province's two main urban areas, as if the rural experience promised by the sugar shack was

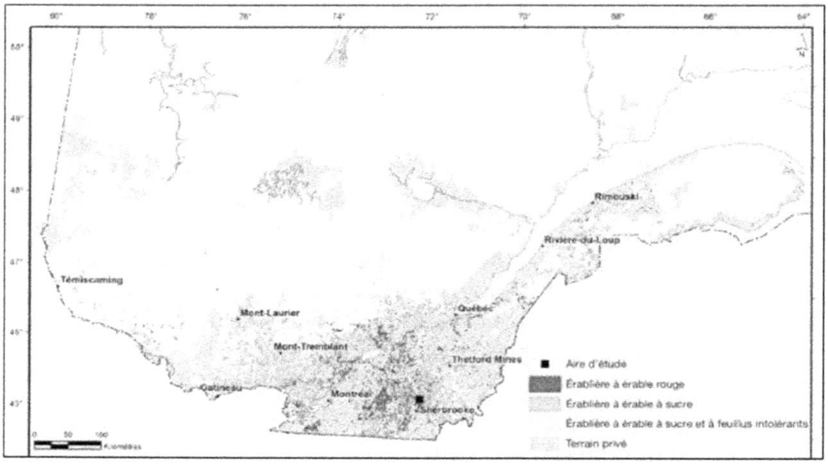

Map 1 Geographic distribution of stands dominated by sugar maple and red maple aged 30 to 50 years in the province of Quebec, according to the mapping of the fourth ecoforest inventory of southern Quebec.

Source: Bédard *et al.* (2018). Effets après 5 ans de l'éclaircie commerciale et de l'amendement du sol sur la production de jeunes érablières en Estrie. 10.13140/RG.2.2.34558.28489.

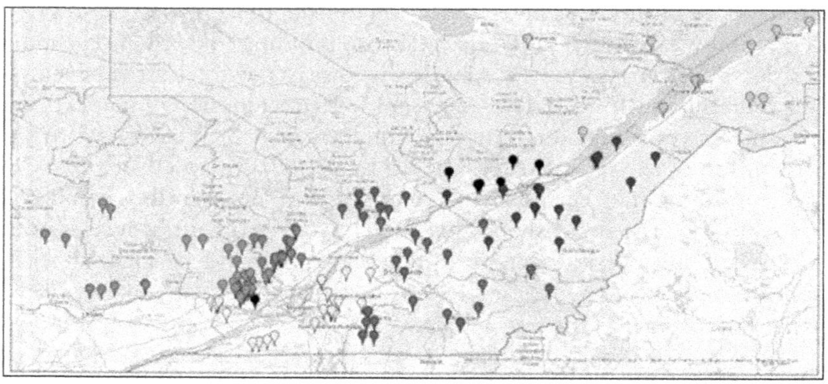

Map 2 Distribution of tourism sugarhouse in the province of Québec

Sources: BonjourQuébec.com & DMO's website. Compiled by Julien Menu 2023

getting ever closer to urbanites while, at the same time, staying at bay from the centers.

A good example of this movement is the Beauce area, south of Québec City. This area is at the core of the global maple production, as it is producing around 50% of the global supply of maple syrup. Despite its importance in the mass production of maple syrup, the Beauce area lists only eight (8) in the number of sugarhouses in tourism listings. On the other hand, the Laurentian area, on the edge of the suburban areas of the Montreal metro, accounts for 29 tourism sugarhouses, most of them along the main provincial roads.

The geography of sugar house tourism and maple syrup production is also framed by multiple intersecting policies and regulations. A commodity in colonial times (Franzen *et al.*, 2018), maple products are now a global commodity filed into international trade agreements, a luxury good for some foreign markets, and a consumption staple for domestic markets. On the ground, it is framed by the agricultural zoning scheme of Québec (Loi sur la protection du territoire et des activités agricoles, 1978), but also through the public forest management plan (Ministère des forêts, 2022), keeping the line of demarcation between the agrarian sugarer and the forest sugarer of Dupont (1975). Also, the actual setting of the leisure and tourism operations are moving it into the food and beverage regulations, with all the sanitary rules that refrain the capacity to keep it in a rustic forest context, a situation that also intersects with seasonality as the expectations in terms of sanitary regulation for a short seasonal activity are similar to the yearlong operation of a restaurant. This last entanglement pushed larger producers towards leaving the food service sector to focus on producing bulk size maple syrup sold through the main producer cooperative. Therefore, these regulations act within the geography as two lines of flights moving apart but still linked by a thread of maple sugar, one as a productive deterritorialized commodity production that reterritorialize along the trade output with a strong expressive content distilled from the territory, and the other as a line of flight, the festive food consumption organized around leisure mobility and entangling multiple contemporary coding (gastronomy, pop culture, traditional imagery, foodies' imaginary). The next section will look at that coding.

A Window on the Tourism Sugarhouse Activities

Looking at the websites of the DMO's listed sugar house, we can see three main ways to recode the sugar shack and the maple trees, as leisure and tourism dimensions: the heritage and tradition coding, the outdoor leisure coding, and lastly, the food and festive coding. One of the main topics of the tourism sugar house website is a reframing of their activities as a continuity of a traditional practice anchoring it in tradition. They use codes from the past to create a form of legitimacy. Those codes are built by adding up images and pictures of metal pails hanging on maple trees, wooden plank shacks, the steam, the red checkered tablecloths, to name just a few. These representations invoke the iconography of the farm sugarer of the proto-industrial era, favoring the metal buckets and stainless evaporators over the vacuum tubing and mechanization it actually uses.

That iconography also creates a reenactment of the visual of the can in which maple syrup is sold in Quebec (Figure 5.1). That particular can is a central point in the contemporary tourism sugar house assemblage in Québec.

First of all, Québec is the only place where the main commercial output of maple syrup is in a metal can, that particular metal can with its iconography. Everywhere else, it is commercialized in bottles. The can, with its emblematic depiction of a maple grove and sugar shack activities, became a visual reference associated with quality and authenticity. One element of the can is the importance of the bulk production and commodity trading of maple syrup in Québec, compared to the other geographies of maple syrup (Lange, 2017). Maple syrup producers in Québec have deterritorialized the traditional farm production to allow large-scale production and commercialization through one large cooperative. This deterritorialization is creating an axis of othering, between tourism maple sugar tradition and production, between the staging of the maple sugar experience, and the maple grove as the sole means of production. The former has values in its accessibility from core urban market, the latter has values through its rent capacity expressed in pounds of syrup per tree. Then, the cooperative resales it, through different forms to supermarkets, restaurants, food transformation businesses or to some tourism sugar houses. The can, with its iconography, is a form of reterritorialization through the expressive material while selling a product that is produced and packaged as a bulk commodity. This line of flight along the desire of affirming the place-based historical importance of this production also allows the tourism sugar house to assemble with it, reproduce its iconography.

The second recoding is along outdoor activities. Sugar season being an in between moment in terms of winter and summer activities, it is depicted as an outdoor moment coupling the tasting of maple products with skating, walking, horse riding, and so on. The maple trees theme are a setting, a backdrop, a part of the landscape allowing for outdoor activities. This coding also brings issues of wellness. While this form of recoding does not remove the traditional code, it completes it and sends the practice into a different materiality that decenters part of the maple food product to give some space back to the maple grove and its forest qualities.

The third, and arguably most important coding, is along the food and festive images. This is central in all the tourism sugar houses, and it is also the one that enables the deterritorializing movement and reassemblage of

Figure 5.1 Maple syrup can
Source: Bulletin des agriculteurs, 24 juin 2022.

the code with other assemblages like take-out sugar house, urban sugar house and so forth. In fact, it is through the food and festive iconography that the tourism sugar house assembles with the remaining of the tourism industry. It is also gastronomy that allows some producers to free themselves from the territorializing movement of seasons and climate and to operate year-long to offer the maple experiences in high tourism seasons like summertime and the winter holiday seasons. While this expressive material silences its links to the traditional French-Canadian Catholic Easter meal, we can still identify common threads of this side of the assemblage, with the presence of eggs, hams and meat pies, that were all forbidden food during the fasting of the Lent.

The coding through food is also centered around the sugar house, reterritorializing the practice inside after deterritorializing the trees and the religious influence of the meal. This reterritorialization along the internal characteristics of the meal entangles with expressive materials recalling elements of tradition and the material conditions of production without being in situ.

From the Sugar Shack to the Tourism Sugar House: Assemblage and *Othering* of the Maple Tree

Coming back to the origins, the maple tree, and its derived form of sugar, was central to the relationship of indigenous people to the world. The centrality was expressed through a cosmogony and kinship with the tree. This was extended in colonial times through a transformation of maple sugar from a ceremonial relationship to a commodity that could be traded (Franzen *et al.*, 2018). This was done through a first deterritorializing movement of the assemblage to be reterritorialized within the colonial trade route. In the St-Lawrence river valley, the trees were deterritorialized to be transformed as a means of production when the first evaporator reterritorialized the production inside the shack. This movement makes invisible the indigenous relationship to maple trees, and indigenous culture as a whole.

This reterritorialization around the sugar shack, where the shack becomes the molar unit of the assemblage, allows the French-Canadian settlers to recode the maple sugar production within their own cultural assemblages. The sugar time became a line of desire restrained by the Catholic heritage as a sugar season was recentered more or less around the fast of Lent before Easter. This moment in time assembled the food code of the breaking of the fast to the sugar shack food of the sugar party, while keeping the line of flight of the deterritorializing of the maple tree active. But still, at that point, the syrup, the sugar and the food were still highly associated with the maple grove and the access to the locations where they were produced.

In the 1950s, a new movement of deterritorialization is sent in movement by the development of production equipment and by the automobiles and tourism associated with the democratization of the cars in North-America by callings for the building of tourism sugarhouses along the roads where tourists are driving to explore the province of Québec. This reterritorialization along tourism mobility, and through urban proximity is still visible these days by the donut-shaped concentration of tourism sugarhouses around the Montreal Metro while the bulk production of maple syrup is moving north and east. This creates a movement of othering as the maple products are stratifying as a code of difference of their cultural assemblage.

This movement brings us today to a situation where the food dimension is the main code of the sugar house tourism, bringing the trees as a mere part of the decoration and surroundings. The sugar maple becomes a means of production while maple products are reterritorialized around the figure of the cabin, where the tree becomes expressive material practically in decoration. This reterritorialization around the cabin is part of the transition to capitalistic valorization and a dynamic of concealing the indigenous heritage. Nevertheless, the maple grove is emerging as a translator of environmental changes and a reterritorialization via proximity, especially as the pandemic limited the capacity of people to gather inside the sugarhouses.

Through all of those movements, maple trees as more-than-human subjects finds its place in this assemblage in constant movement. It moves from kin in indigenous cosmogony to a means of production in a rural economy to a part of the scenery in a tourism sugar house centred on staged heritage and food and festivities recoding. This movement of othering of the trees is followed by the invisibilization of indigenous legacy and culture, othering through cultural and tourism niche products through sugar seasons as immaterial heritage and, finally, a full othering of the maple tree.

This example of movement of deterritorialization/reterritorialization in tourism is not solely relevant to understand the transformation of immaterial heritage like the sugar time tradition in Québec. It is a demonstration of the capacity of tourism to disconnect phenomenon from the surrounding, to Other it through commodification along the line of desire of the tourist mobile subjectivity (Lapointe, 2020, 2024). This othering process of tourism implies a double bind in which tourism support and reproduce heritage, culture, places, but also commodify it in ways to make it available to the line of desires of tourist's consumption. Our research stresses the importance of paying attention to those movements of deterritorialization and how they reterritorialize. While the reterritorialization has an implicit spatial dimension, the movement of transformation of sugar shacking exposed through our investigation underlines that it is also how it recombines the expressive dimension of the previous assemblage, in our case making it possible to move the experience closer to the consuming tourists, less and less connected to the ecological conditions of the maple trees, while opening up large scale production of bulk maple syrup with low entanglements with tourism.

Conclusions: Reassembling Tourism Futures in a Changing Climate

As we finalize this chapter, Québec is experiencing its warmest winter in 150 years, with an early spring disrupting maple syrup production. While the reterritorialization of tourism around food-based experiences remains stable for now, climate change repositions the maple tree as an actant (Callon, 1984; Latour, 2005), signaling new trajectories for tourism and heritage futures. Drawing from Bergman *et al.* (2010), two contrasting scenarios emerged: a dystopian future scenario and a utopian one.

Dystopian scenario

In a dystopian future, maple syrup production becomes increasingly detached from its ecological roots. Rising temperatures push the maple sugar belt further north, marginalizing traditional sugar shacks and severing ties between tourism and the maple forest. In this scenario, sugarhouses become hyper-commercialized, urban-centered experiences, offering a nostalgic but artificial re-creation of past traditions. Maple syrup, now a mass-produced global commodity, is divorced from its terroir, with tourism merely simulating a lost connection to nature.

Utopian scenario

A utopian future envisions a reterritorialization of maple tourism that acknowledges its ecological entanglements. Here, the sugar maple is not just a raw material but a companion species (Haraway, 2016), fostering new ecological and cultural relationships. Tourism evolves into an immersive, place-based experience, where visitors engage with the living forest, the climate crisis, and the sociocultural histories that shape maple syrup production. Sugarhouses become spaces for ecological education and climate adaptation, challenging the capitalist logic of tourist consumption.

To navigate between these futures, we turn to Assemblage Theory (Deleuze & Guattari, 1980) and Inayatullah's (1998) CLA. Assemblage Theory helps us trace the shifting material, cultural and economic entanglements that shape maple tourism, revealing its inherent instability and potential for reconfiguration. Meanwhile, CLA enables a deeper excavation of the layered realities structuring these futures – from surface-level consumer trends to systemic economic shifts, dominant narratives of authenticity and heritage and foundational myths of human-nature relations (Inayatullah, 1998). By integrating these approaches, we move beyond linear projections and open new possibilities for tourism futures that embrace complexity, relationality and adaptation.

The future of maple tourism is not predetermined. Rather, it is an evolving assemblage, constantly deterritorializing and reterritorializing. Whether it trends toward dystopia or utopia depends on the sociopolitical choices and material practices that shape its unfolding reality.

Notes

(1) Le Cercle des fermières, is still today the largest association of rural women in Québec.
(2) For the remaining of the chapter we will use the terms Sugarhouse to refers to the tourism-based maple sugar sites, and sugar shack for the artisanal site of productions.
(3) Free translation by the authors.

References

Ateljevic, I., Pritchard, A. and Morgan, N. (eds) (2007) *The Critical Turn in Tourism Studies: Innovative Research Methodologies*. Elsevier.

Bédard, S. and Gauthier, M.-M. and Guillemette, F. and Ouimet, R. (2018) Effets Coaprès 5 ans de l'éclaircie commerciale et de l'amendement du sol sur la production de jeunes érablières en Estrie.

Bergman, A., Karlsson J.C. and Axelsson, J. (2010) Truths claims and explanatory claims – an ontological typology of futures studies. *Futures* 42 (8), 857–865.

Braidotti, R. (2011) *Nomadic Theory*. Columbia University Press.

Braidotti, R. (2019) *Posthuman Knowledge*. Polity Press.

Briassoulis, H. (2017) Tourism destinations as multiplicities: The view from Assemblage Thinking. *International Journal of Tourism Research* 19 (3), 304–317.

Callon, M. (1984) Some elements of a sociology of translation: Domestication of the scallops and the fishermen of St Brieuc Bay. *The Sociological Review* 32 (1), 196–233.

Chamberlain, A.F. (1891a) The maple amongst the Algonkian tribes. *American Anthropologist* 4 (1), 39–44.

Chamberlain, A.F. (1891b) Nanibozhu amongst the Otchipwe, Mississagas, and other Algonkian tribes. *The Journal of American Folklore* 4 (14), 193–213.

Chapais, J.C. (1883) *Guide Illustré du Sylviculteur Canadien*. Senécal.

Deleuze, G. and Guattari, F. (1980) *Mille Plateaux*. Editions de Minuit.

Dovey, K. (2009) *Becoming Places: Urbanism/Architecture/Identity/Power*. Routledge.

Dupont, J.-C. (1975) *Le Sucre du Pays*. Leméac.

Dupont, J.-C. (2004) *Le Temps Des Sucres*. Les éditions GID.

Ek, R. and Tesfahuney, M. (2016) The paradigmatic tourist. In A.M. Munar and T. Jamal (eds) *Tourism Research Paradigms: Critical and Emergent Knowledges* (Vol. 22, pp. 113–129) Emerald Group Publishing. https://doi.org/10.1108/S1571-504320150000022013.

Équipe MIT (2002) Tourisme 1: Lieux communs, coll. *Mappemonde*. Belin.

Extension, C.C. (2019, July 26, 2019) Maple Syrup Season – What causes the sap to flow out of the trees? Retrieved from https://monroe.cce.cornell.edu/agriculture/seasonal-produce-highlights/maple-syrup-season-what-causes-the-sap-to-flow-out-of-the-trees

Franklin, A. (2012) The choreography of a mobile world: Tourism orderings: Adrian Franklin. In R. Van der Duim, C. Ren and G.T. Johannesson (eds) *Actor-Network Theory and Tourism* (pp. 58–73). Routledge.

Franzen, J.G., Martin, T.J. and Drake, E.C. (2018) Sucreries and Ziizbaakdokaanan: Racialization, Indigenous creolization, and the archaeology of maple-sugar camps in Northern Michigan. *Historical Archaeology* 52 (1), 164–196.

Friedl, E. (1956) Persistence in Chippewa culture and personality. *American Anthropologist* 58 (5), 814–825.

Gaudreau, V. (2020) Plaisir intemporel. *Continuité* 164, 32–34.

Haraway, D.J. (2016) *Staying with the Trouble: Making kin in the Chthulucene*. Duke University Press.

Henshaw, H.W. (1890) Indian origin of maple sugar. *American Anthropologist* 3 (4), 341–352.

Höckert, E. (2018) *Negotiating Hospitality: Ethics of Tourism Development in the Nicaraguan Highlands*. Routledge.

Inayatullah, S. (1998) Causal layered analysis: Poststructuralism as method. *Futures* 30 (8), 815–829.

Kimmerer, R. (2013) *Braiding Sweetgrass: Indigenous Wisdom, Scientific Knowledge and the Teachings of Plants*. Milkweed editions.

Lange, M. (2017) *Meanings of Maple: An Ethnography of Sugaring*. University of Arkansas Press.

Lapointe, D. (2020) Reconnecting tourism after COVID-19: The paradox of alterity in tourism areas. *Tourism Geographies* 22 (3), 633–638. https://doi.org/10.1080/14616688.2020.1762115.

Lapointe, D. (2021) Tourism territory/territoire (s) touristique (s): When mobility challenges the concept. In M. Stock (ed. *Prog) ress in French Tourism Geographies* (pp. 105–116). Springer.

Lapointe, D. (2024) Alterity, mobility and territory: Conceptualising tourism space in a world in flux. In J. Wilson and D. Muëller (eds) *Handbook of Tourism Geographies* (2nd edn). Routledge.

Lapointe, D. and Coulter, M. (2020) Place, labor, and (im)mobilities: Tourism and biopolitics. *Tourism Culture & Communication* 20 (2–3), 95–105.

Lapointe, D. and Muldoon, M. (2023) Tourism entanglements: Complexity and multiplicity at play. *Tourism Culture & Communication* 24 (2), 89–94. https://doi.org/10.3727/109830423X16969739300601

Latour, B. (2005) An introduction to actor-network-theory. *Reassembling the Social.*

Loi sur la protection du territoire et des activités agricoles (1978) Gouvernement du Québec, § 27, p-41.1.

Méthot, J.-R. and Rompré, N. (1952) *L'érable à sucre du Québec*. Gouvernement du Québec.

Ministère des forêts, d.l.F.e.d.P. (2022) *Plan directeur ministériel pour le développement de l'acériculture en forêt publique.* Gouvernement du Québec Retrieved from https://mffp.gouv.qc.ca/wp-content/uploads/PL_directeur_acericulture_consultation_MFFP.pdf

Novelli, M., Cheer, J.M., Dolezal, C., Jones, A. and Milano, C. (2022) *Handbook of Niche Tourism*. Edward Elgar Publishing.

Picard, D. and Di Giovine, M.A. (2014) Introduction: Through other worlds. *Tourism and the Power of Otherness: Seductions of Difference* (pp. 1–28). Channel View Publications.

Sulte, B. (1918) *Mélanges Historiques*. G. Ducharme.

Part 2

Navigating Spatial Entanglements with the 'Other' and Implications for Tourism Futures

6 The Future of Rural Tourism and Leisure Literature: Perpetuating or Challenging the *Othering* of Farm Animals?

Giovanna Bertella and Ludovica Lombardini

Introduction

Animals used in farm-related activities such as food production and agricultural work ('farm animals') are the ultimate 'others'. That is, they are considered to belong to what Cohen (2019) referred to as the 'absolute "Other", an ontological category fundamentally different from humans' (2019: 419). Farm animals are exploited by farmers to gain profit, used by recreationists and tourists to gain memorable experiences, and overseen by most leisure and tourism scholars. This study focuses on the latter: the lack of engagement in farm animal issues by scholars in the leisure and tourism fields. This study's ambition is to provoke critical reflection among scholars, especially those who, despite their engagement in rural or even animal issues, pay little or no attention to farm animals. Indeed, such neglect has the unfortunate consequence of contributing to a process of the 'othering' of beings whose sentience (i.e. the capacity to have feelings and be aware of a variety of states and sensations) is scientifically recognised although not fully understood (Proctor, 2012). The research question is: what space will farm animals occupy in the future scholarly literature about tourism and leisure?

This chapter starts with a review of the leisure and tourism literature about farm animals and presents the concept of the *lives of the rural*, which is one of the three dimensions of the rural space model proposed by Halfacree (2006) and was adapted to apply to animals (Halfacree & Williams, 2021). This spatial dimension is viewed as an arena, which is a metaphorical place where ideas, arguments, and positions about farm animals are challenged and, ultimately, the identity of animals is negotiated and emerges. We then describe this study's methodology, which relies on the use of dialogue between the authors: one being an academic and the second a practitioner with first-person experience with farm animals and tourists and a vivid engagement in animal well-being from which derived the establishment of *La Tana del Bianconiglio*, an agrotourism and animal sanctuary in Tuscany (Italy). We analysed animal–tourist encounters using the metaphor of scholars as tourists and explored two scenarios involving alternative futures for the tourism and leisure scholarly literature in relation to farm animals. This is done by drawing on Bergman *et al.*'s (2010) typologies of future studies to identify two opposite scenarios: a dystopian future scenario that perpetuates othering, and a utopian future scenario that challenges othering. Hence, by focusing on the under-investigated topic of farm animals as 'others', our study contributes to providing more critical tourism and leisure literature for future tourism research. Methodologically, this study responds to the call by scholars who have advocated for novel approaches and are able to capture, to a certain extent, the perspectives of animals (Bertella, 2022; Chakraborty, 2021; Cohen, 2019; Guia & Jamal, 2020; Tomassini & Bertella, 2023; Tomassini *et al.*, 2022).

Farm Animals in Tourism and Leisure

Although many farm animals live in the countryside, most studies about rural tourism and leisure do not discuss them. For example, a recent study about rural tourism focused on livestock operations and factory farming and addressed various perspectives of the rural landscape (Caffyn, 2022). The study hardly mentions the presence of animals, which are predominantly referred to as disturbing elements in the landscape. Borrowing an expression by Nixon (2011) who discussed 'slowly unfolding environmental catastrophes' (2011: 2), this disturbance is reported by Caffyn (2022) as a form of 'slow violence', which is explained as a sensorial disruption to the idyllic countryside landscape as it leads to a view that does not correspond with tourists' expectations and causes unpleasant smells and increased noise. The animals, whose bodies are restrained in the discussed farming facilities and constitute the sources of such 'slow violence', are ignored. In this chapter, we explore how scholars who discuss tourism and leisure in the rural space tend to focus on certain aspects, such as the landscape, gastronomy (e.g. Bessière, 2013; Sims, 2009) and events (e.g. Lillywhite *et al.*, 2013; Von Essen *et al.*, 2020), but overlook the animals' presence or consider it a means to human enjoyment. In the reported example (Caffyn, 2022), the existence of the animals is not completely ignored, but it is presented uncritically as a disturbance to tourism and leisure activities instead of a brutal form of violence towards sentient beings.

The limited attention paid to farm animals also characterises tourism and leisure studies about animals, most of which concern wildlife, followed by animal companions (especially dogs) and equines (mainly horses). The most recent studies concerning animal welfare and ethics published in well-established journals have almost entirely ignored farm animals. For example, among the studies from the recently launched *Annals of Tourism Research* collection about animal ethics (*Curated Collection: Animal Ethics in Tourism*), only a few touch on the circumstances of farm animals as part of the broader discussion about animals in tourism (e.g. Fennell & Thomsen, 2021; Winter, 2020) and in relation to food (Bertella, 2020). Recently, the *Journal of Sustainable Tourism* published a special issue (2023 Special Issue: Sustainability Dimensions of Wildlife Tourism: Welfare, conservation, practices, and impacts, Vol. 31, issue 3) and the *Journal of Ecotourism* started a collection (*Collection: Wildlife Tourism*) about animals, but the focus of both is on wildlife. What about the many sheep inhabiting the countryside that tourists visit? The countless hens and pigs from which the ingredients for visitors' meals come? The cows that graze in the fields while tourists enjoy taking pictures of them?

Despite such scarce interest by scholars in farm animals, some valuable contributions can be found in the form of chapters by a few engaged scholars in edited books. For example, two books about animals and animal consumption in tourism edited by Carol Kline (2018a, 2018b) deepen the topic of the use of animals as food in tourism, which is partly discussed by Yudina and Fennell (2013). For example, the chapter by Venegas and López (2018) reflects critically on the tradition of eating newborn goats. A few other contributions about farm animals can be found in a handbook about tourism and rural community development (Kline, 2023) and a book about animals as workers in the tourism industry (Rickly & Kline, 2021). Among the chapters in the latter, Tully and Carr (2021) writes about the proper respect that farm animals in the leisure and tourism sectors deserve. This echoes the call by Neil Carr (2009) in the editorial of the special issue of *Current Issues in Tourism* on animals in tourism and leisure and the reflections in his edited book about domesticated animals (Carr, 2016).

Such short review shows the extremely limited number of tourism and leisure scholars who pay attention to farm animals. The first author of this chapter has contributed to the literature via selected studies (Bertella, 2018a, 2018b, 2021, 2023; Tomassini & Bertella, 2023). One such work discusses farm animals in rural tourism in relation to justice by attempting to reveal the underlying logic of power that is common to marginalised groups and silences some among them (Bertella, 2023). Another noteworthy study explores the experience of a rural event from the point of view of one of the animals involved (Bertella, 2021). This study notes the paucity of critical leisure and tourism studies about farm animals and the perpetuation of the sharp dualism between humans and others (nonhuman animals). The study concludes by asking why so few tourism and leisure scholars engage in discussions about farm animals. The current study builds on this question and aims to provoke reflections about the future of farm animals in the scholarly literature.

Farm Animals in the Rural Space

The way farm animals are addressed in the scholarly literature on the rural space has changed considerably over the last few decades. As observed by Cloke (1989), in the 1970s, some geographers and sociologists started to direct their attention to the animals inhabiting the rural space and this branch of studies has grown considerably since (e.g. Jones, 2003, 2006; Philo & Wilbert, 2000; Tovey, 2003). These developments favoured the emergence of rural animal studies, a field of study that reflects on the concepts of space (understood both in its spatial dimension and metaphorically), environmental issues and animal welfare and rights (Buller & Morris, 2003; Jones, 2006). Once considered mere units of production, farm animals have been studied by scholars who consider them as central agents in the rural space. Controversial questions have been posed about agricultural practices, including those concerning the relations between animals and their handlers (Buddle *et al.*, 2021; Gorman, 2018; Holloway *et al.*, 2009; Jones, 2003).

This change is relevant to the current study because relations and encounters are central to tourism and leisure, and the recognition of animals as agents thus allows for reflection on the possible processes related to their 'othering'. In this context, a useful model about the rural space is that proposed by Halfacree and Williams (2021). These scholars adapted the rural space model by Halfacree (2006) to apply to animals and invited researchers to 'give greater attention to how the spaces we define as rural are co-produced by the more-than-human, not least other animals' (Halfacree & Williams, 2021: 375). This model understands the rural space through three dimensions – *locality*, *representations of the rural*, and *lives of the rural* – and Halfacree and Williams (2021) argued for including the animals' narratives in this heterogeneous assemblage. Exploring the *lives of the rural* dimension and following Jones (2013), Halfacree and Williams (2021) reflect on the bonds that tie humans and animals together. For example, they discussed the lives of sheep as part of the sheep-shaping of British rurality. Their lives are described as developing through autonomous activities that are not structured and are controlled by humans, life events (e.g. birth, experiences of diseases and pregnancies) as well as physical, emotional and affective interactions with humans, mainly farmers.

Based on these considerations and to answer the research question about the space of farm animals in future scholarly literature about tourism and leisure, we focused on the *lives of the rural* dimension, with particular attention to animal–tourist interactions. The main idea is that this dimension constitutes the arena in which the identity of the farm animal is negotiated through different performances, reflections and feelings by all the involved parties. This arena is decisive in terms of the attitudes and behaviours characterising human–animal encounters and ultimately the possible 'othering' processes. To explore the *lives of the rural* dimension and with the aim of shifting the focus from an exclusively human perspective towards a relational one that includes the perspective of the 'other', we adopted a methodology based on the use of dialogue and the metaphor of researchers as tourists.

Methodology

Our methodological choice concerning the use of dialogue stemmed from three considerations. The first was about dialogical methodology being based on the belief that meanings can be transformed through the research act of individuals who have multiple interconnected intentions, among which the one to 'explore and articulate personal and collective narratives of resistance in relation to dominant discourses and metanarratives' (Norris & Sawyer, 2016: 9). These characteristics resonated with our aim to critically examine the process of 'othering' experienced by farm animals. The second consideration concerned the fact that we were influenced by the recognition among scholars that dialogue offers a valuable and democratic avenue for exploring unfamiliar realities, facilitating a genuinely collaborative approach to knowledge production (Paulus *et al.*, 2008; Chen *et al.*, 2013; Sinha & Back, 2014). This perspective was particularly pertinent to our study, given that much of our data originated from the experiences of the second author, who is the co-owner and manager of an agrotourism business that functions as a farm animal sanctuary – *La Tana del Bianconiglio* in Tuscany (Italy) – as well as the main caregiver of the animals.

The third consideration influencing our methodological choice pertains to the use of dialogue, which serves to illuminate the emergent, nonlinear nature of the research process. Data are generated and collectively interpreted along an ongoing conversational trajectory, resulting in a 'fluid text' that affords readers the opportunity to observe the evolution of meanings (Norris & Sawyer, 2016). This characteristic underscores a significant reader engagement, which we deemed essential for fostering profound reflections on a frequently overlooked and occasionally contentious subject matter.

To explore what space farm animals will occupy in the future scholarly literature about tourism and leisure, we focused on the *lives of the rural* and directed our attention to the relational dimension of rural tourism, and more precisely, the encounters between animals and tourists. We strived to explore and represent both sides of the encounters. With respect to the human side, the first author had a first-person experience as she paid a visit to the animals at the sanctuary in January 2022, and the second author has witnessed many animal–tourist encounters since starting the business in 2016. From the animal side, we collaborated closely and the second author's role as animal caregiver was essential: her first-person daily experience with animals gave her the credibility to represent animal–tourist encounters as well as, at least at a certain extent, the perspective of animals in these encounters. We recognise the limitations inherent in articulating the viewpoints of other beings, especially those outside the human species, yet we contend that the perspective we present holds a degree of plausibility (Vickers, 2010; DeMello, 2013).

The data about the farm animal–tourist encounters were collected during a series of informal conversations between the two authors and are presented in the following section in the form of a dialogue. Such conversations, along with email exchanges, occurred online in April–June 2023 and included the revision of the draft of this book chapter, especially the parts dedicated to the findings, discussion and conclusion. As aforementioned, the use of dialogue is intended to show the reasoning behind the identification of the typologies of the encounters and tourists. To add dynamism to the text in line with the advice of Halfacree and Williams (2021), the dialogue is enriched with pictures provided by the second author.

About the use of the metaphor, the gained insights were analysed by drawing a parallel between the typologies of tourists who encounter farm animals and scholars who study rural tourism. The choice to use the metaphor of scholars as tourists was based on the consideration of the use of metaphors as a valuable way to understand and provide orientation to researchers in their scholarly endeavours and to promote self-reflection (Lakoff & Johnson, 2008; Schmitt, 2005). Rathmayr (1991), as reported by Schmitt (2005), described researchers engaged in qualitative studies using metaphors, for example, 'hunters' and 'hikers'. Similarly, the current study used a metaphor to describe the scholars by comparing the various typologies of tourists that the second author meets during her job with scholars researching rural tourism and leisure. Within the tourism literature, Stinson *et al.* (2021) discussed researchers' practice of (metaphorically) becoming common plantain, pervasive and rooted in the object of study, and argued how researchers' knowledge and power can enhance justice by research that is both 'productive and disruptive force of change' (2021: 247). Likewise, our metaphor of scholars as rural tourists is intended to illustrate scholars' role in relation to the 'othering' processes towards farm animals. This metaphor and the related reflections form the basis of two future scenarios inspired by Bergman *et al.* (2010) typologies of future studies, representing two alternative futures: a dystopian and a utopian one – for the space of farm animals in the tourism and leisure literature.

Farm Animal–Tourist Encounters

Author 2: I have witnessed many encounters between tourists and the animals living at our sanctuary. Actually ... you were one of these tourists when you visited us last year. How was the experience for you?

Author 1: You're right. Let me think ... before the visit, I had read about some of the animals on the web page ... I was curious to meet Senia. I knew about her jumping out of the truck on the way to the slaughterhouse and making the headlines of the local newspapers.

Author 2: Her story attracted a lot of attention ... also among our tourists: everyone likes to hear about Senia's escape. Some tourists pet Senia on her snout and reassure her that now she's in a safe place.

Author 1: Are many of your tourists vegan or vegetarian?

Author 2: Some are but not all. Senia is liked by everybody. I see a sharp conflict between the tourists acting in the way I told you with Senia and then going to the nearby restaurant and eating the local ham. Such behaviour is very common. When we posted about Senia's story on social media just after her rescue, we chose to include a picture of the truck that was transporting her and the other pigs (Figure 6.1). Sometimes the tourists meet animals in this way: the animals inside trucks and the tourists in their cars passing the truck. The animals are probably too scared to be aware of these encounters, and the tourists ... they don't really seem to pay attention to the animals inside the trucks ... they may see parts of their bodies in-between the bars, but ... they turn away ...

Figure 6.1 The animal transport truck from which Senia escaped
Source: Ima Assiak/ Wisteria Ima Assiak

Author 1: How do you explain the tourists' behaviour? I mean ... ignoring what is pretty much in front of their eyes?

Author 2: I think that some people pet the pig because it's a kind of fun and unusual ... They take pictures and even selfies with the pigs ... and then ... most of them eat pork and ham ... because they like the taste of it ... The tourists also want to try typical local products ... The encounter with the animals is only an episode ... then ... back to their habits and to what is promoted as a gastronomic experience. After petting a pig ... the next encounter with a pig is at the table of the local restaurant ... .

Author 1: Thinking back on my experience, meeting the animals made an impression on me. I have been interested in animal issues for a while. Hearing stories about their short lives and the killing of most farm animals, not to mention the stories of neglect by some farmers, and then meeting the animals, the ones living at the agrotourism sanctuary ... the few lucky ones ... it made everything much more real to me.

Author 2: Some tourists feel like you and get somehow attached to the animals. I'm pretty sure that there is no really strong bond from the animals' side, maybe some curiosity ... we are talking about short encounters (Figure 6.2). Tourists are not the ones taking care of the animals ... every single day ... feeding them, checking on their health and so on. The encounters with the tourists are only a small part of the animals' lives: the rest of their lives is about relaxing, sleeping, eating ... normal pig and sheep things ... But even these short encounters can change the lives of the

Figure 6.2 A tourist paying a visit to the animals at the agrotourism sanctuary
Source: Ima Assiak/ Wisteria Ima Assiak

tourists: they can gain some first-hand experience with farm animals, which is rare for urban people, and maybe become more sensitive to animal issues. I always hope that the visit to the sanctuary can plant some seeds of change. And these encounters can be important for the animals, especially when the tourists end up adopting an animal and contributing to the expenses for her care. Sometimes we have guests who are engaged activists. We had a visit from two journalists who were interested in vegan travels … they took many pictures of the animals and dedicated a blog post to our agrotourism sanctuary.

Author 1: I imagine that you have some families visiting the agrotourism sanctuary. How do children react to the animals?

Author 2: It depends. Most children are curious. Some children want to pet and play with the animals as they would do with dogs, and others want to help by feeding them or cleaning the barn. Certainly, some are not afraid to get their hands dirty! These are sometimes children from families where the parents are sensitive to animal issues, but not always … Sometimes, no matter what their parents say or do, the children do their best to help the animals!

Discussion: Dystopian and Utopian Scenarios of Farm Animals in the Future of Tourism Literature

To discuss the space of farm animals in the future scholarly literature about tourism and leisure, we used the range of animal–tourist encounters that emerged from the data of the dialogue reported in the previous section as a point of departure and then focused on the typologies of tourists. We used the metaphor of scholars as tourists to reflect on extant and future scholars' engagement with farm animal issues in tourism and leisure. This process led to the identification of two scenarios, namely a dystopian and a utopian one (Bergman *et al.*, 2010), which are presented in the following sections.

Perpetuating Othering: A dystopian scenario

We used the three types of encounters discussed in our dialogue to develop the *Perpetuating Othering* scenario. The first type of encounter was that between the tourists and animals during the transport of the latter from farms to slaughterhouses. These encounters are mainly visual, and none of the parties is in a position or willing to develop the encounter into a mutual relationship. It is plausible to argue that the animals are in distress and unable to pay attention to the tourists. It can be assumed that the tourists do not proactively seek to view of the animals. This behaviour could be derived from a lack of interest and/or willingness to witness how the animals are transported on trucks, presumably in a state of fear as in the case of Senia, or why, namely, to be slaughtered and become food like that served in the restaurants close to the agrotourism sanctuary. The second type of encounter we used to develop this scenario occurs at restaurants and concerns the tourists consuming animal flesh. Obviously, this encounter excludes the emergence of any relationship as it consists of the final annihilation of the animals, where they are reduced to objects of sensorial pleasure and symbolic value by the tourists.

The behaviour of the tourists in these two types of encounters aligns with what was observed in the theoretical part of this chapter about the extant scholarly literature on rural tourism and leisure. This literature, exemplified by studies on the rural landscape and gastronomic heritage (e.g. Bessière, 2013; Caffyn, 2022; Sims, 2009), explores rural destinations without delving into the possible problematic issues concerning the animal inhabitants of the destinations. In the same way as the tourists passing by animal transport trucks look away from the enclosed animals, and those enjoying ham at the local restaurant seem to forget about the pig that they just petted, some scholars investigate the rural world, but only partially, because they exclude some of the actors who contribute to making the world as it is (Halfacree & Williams, 2021).

Similar considerations can be made about the third type of farm animal–tourist encounter that emerged from the data. In this encounter, the tourists are temporally fascinated and/or entertained by the animals. The dialogue makes references to the tourists, both adults and children, enjoying petting and taking pictures of the animals. A parallel can again be drawn between tourists visiting the countryside and researchers investigating rural tourism and leisure. This type of encounter, which is occasional and rather uncritical, resembles some of the contributions from the extant literature, for example, in the case of attention being paid to farm animals, even in positive terms, but only within the exploration of rural events (e.g. Lillywhite *et al.*, 2013; Von Essen *et al.*, 2020). As tourists do not develop any relationship with the animals, some scholars include farm animals in their investigations of rural tourism and leisure but do not engage, at least explicitly, in any deep reflection about the sentience of the animals and its consequences.

Based on these three types of tourist encounters and our reflections using the tourist–scholar metaphor, the *Perpetuating Othering* scenario can be qualified as conservative. The scholars investigating rural tourism and leisure keep their academic attention on topics, in some cases already widely discussed, that virtually ignore farm animal issues. In the future, there will be no space for farm animals in the scholarly literature. Since they are explored only marginally and rather superficially, farm animals will continue to be the object of a detached and uncritical process of 'othering' by scholars.

Challenging Othering: A utopian scenario

Two types of encounters underpin the second scenario, and both concern practical help that tourists offer farm animals. The first involves tourists adopting the animals or using their professions as journalists or academics (as illustrated in the dialogue) to tell the stories of the animals with the intention of informing and provoking reflections in others. Such interactions, although limited in time, can be significant for both parties (i.e. tourists and animals) and have long-term effects. This type of encounter can be associated with the work of the few scholars who have adopted the cause of farm animals (e.g. Bertella, 2023; Kline, 2018a, 2018b; Tully & Carr, 2021) and whose contributions can make these animals more visible to other academics. The second type of encounter in which the second scenario is based concerns the children who visit the agrotourism sanctuary and, as reported in the dialogue, are not afraid to 'get their hands dirty'. They sometimes mirror their parents' attitudes and behaviours and sometimes interact with the animals regardless of what their parents say.

Using the academic–tourist metaphor, this scenario points to a future where a new place will exist for farm animals in the scholarly literature. In the future, farm animals will be studied to the same extent as pets and wildlife. Farm animals will be explored as both charismatic and familiar beings. Importantly, this new literature will be in line with scientific progress and ethical considerations about animals, farm animals included, and will be courageous.

Conclusion

In this chapter, we discussed farm animal–tourist encounters. We highlighted the relational space between rural tourists and farm animals and used the metaphor of scholars as tourists to develop two scenarios about the space of farm animals in future tourism and leisure literature. These scenarios were elaborated on with the point of departure as different types of farm animal–tourist encounters, which were observed and discussed between the authors, one of whom is an academic and the other a practitioner. A possible scenario is that the future literature will be very similar to the present: farm animals will have a marginal position in the literature, and scholars will include them in their studies as a means to various tourism and leisure activities, such as entertainment and gastronomy. The second scenario concerns scholars broadening and deepening their field of interest and applying a critical lens to the way farm animals live and experience tourist encounters in the rural space. In this scenario, the literature about animal issues in tourism and leisure will be more inclusive and innovative, with new and renewed academic voices challenging the status quo.

We would like to close our chapter by inviting readers to reflect on the types of tourists we have presented and to challenge themselves to emulate those children who, are eager to help and irrespective of what their parents say, engage with the 'other'.

References

Bergman, A., Karlsson J.C. and Axelsson, J. (2010) Truths claims and explanatory claims – an ontological typology of futures studies. *Futures* 42 (8), 857–865.

Bertella, G. (2018a) Animals off the menu: How animals enter the vegan food experience. In C. Kline (ed.) *Animals, Food, and Tourism* (pp. 67–81). Routledge.

Bertella, G. (2018b) Vegetarian ecofeminism in tourism: Emerging tourism practices by institutional entrepreneurs. In B.S. Grimwood, K. Caton and L. Cooke (eds) *New Moral Natures in Tourism* (pp. 86–98). Routledge.

Bertella, G. (2020) Re-thinking sustainability and food in tourism. *Annals of Tourism Research* 84, 103005._

Bertella, G. (2021) Animal-based experiences and animal experiences: Farm animals' perspective on human leisure in rural settings. *Annals of Leisure Research* 24 (5), 631–645.

Bertella, G. (2022) Including the 'voices' of animals in tourism, hospitality, and leisure research. In F. Okumus, S.M.S. Rasoolimanesh and S. Jahani (eds) *Advanced Research Methods in Hospitality and Tourism* (pp. 199–211). Emerald.

Bertella, G. (2023) Sustainability and justice: Farm animals in rural tourism and lessons to be learned. *Journal of Qualitative Research in Tourism* 4, 1–18.

Bessière, J. (2013) 'Heritagisation', a challenge for tourism promotion and regional development: An example of food heritage. *Journal of Heritage Tourism* 8 (4), 275–291.

Buddle, E.A., Bray, H.J. and Ankeny, R.A. (2021) "Of course we care!": A qualitative exploration of Australian livestock producers' understandings of farm animal welfare issues. *Journal of Rural Studies* 83, 50–59.

Buller, H. and Morris, C. (2003) Farm animal welfare: A new repertoire of nature-society relations or modernism re-embedded? *Sociologia Ruralis* 43 (3), 216–237.

Caffyn, A. (2022) Contested rural landscapes: Contemporary entanglements of tourism and farming. *Journal of Sustainable Tourism* 32 (2), 259–281.

Carr, N. (2009) Animals in the tourism and leisure experience. *Current Issues in Tourism* 12 (5–6), 409–411.

Carr, N. (2016) *Domestic Animals and Leisure*. Springer.

Chakraborty, A. (2021) Emplacing non-human voices in tourism research: The role of dissensus as a qualitative method. *Tourism Geographies* 23 (1–2), 118–143.

Chen, C.Y., Jim Wu, Y.C. and Wu, W.H. (2013) A sustainable collaborative research dialogue between practitioners and academics. *Management Decision* 51 (3), 566–593.

Cloke, P. (1989) Rural geography and political economy. In R. Peet and N. Thrift (eds) *New Models in Geography, Volume 1* (pp. 164–197). Hunwin Hyman.

Cohen, E. (2019) Posthumanism and tourism. *Tourism Review* 74 (3), 416–427.

DeMello, M. (2013) *Speaking for Animals: Animal Autobiographical Writing*. Routledge.

Fennell, D.A. and Thomsen, B. (2021) Tourism & animal suffering: Mapping the future. *Annals of Tourism Research* 91, 103317.

Gorman, R. (2018) Human-livestock relationships and community supported agriculture (CSA) in the UK. *Journal of Rural Studies* 61, 175–183.

Guia, J. and Jamal, T. (2020) A (Deleuzian) posthumanist paradigm for tourism research. *Annals of Tourism Research* 84, 102982.

Halfacree, K. (2006) Rural space: Constructing a three-fold architecture. In P. Cloke, T. Mardsen and P. Mooney (eds) *Handbook of Rural Studies* (pp. 44–62). Sage.

Halfacree, K. and Williams, F. (2021) Advancing rural as 'something more than a human estate': Exploring UK sheep-shaping. *Journal of Rural Studies* 87, 375–385.

Holloway, L., Morris, C., Gilna, B. and Gibbs, D. (2009) Biopower, genetics and livestock breeding: (Re)constituting animal populations and heterogeneous biosocial collectivities. *Transactions of the Institute of British Geographers* 34 (3), 394–407.

Jones, O. (2003) 'The restraint of beasts': Rurality, animality, actor network theory and dwelling. In P. Cloke (ed.) *Country Visions* (pp. 450–487). Pearson Education.
Jones, O. (2006) Non-human rural studies. In P. Cloke, T. Mardsen and P. Mooney (eds) *Handbook of Rural Studies* (pp. 185–200). Sage.
Jones, O. (2013) "Who milks the cows at Maesgwyn?" The animality of UK rural landscapes in affective registers. *Landscape Research* 38 (4), 421–442.
Kline, C. (2018a) *Animals, Food, and Tourism*. Routledge.
Kline, C. (2018b) *Tourism Experiences and Animal Consumption: Contested Values, Morality and Ethics*. Routledge.
Kline, C. (2023) Centering animals within rural tourism. In H. Mair (ed.) *Handbook on Tourism and Rural Community Development* (pp. 344–358). Elgar.
Lillywhite, J.M., Simonsen, J.E. and Acharya, R.N. (2013) Designing a better fair: How important are the animals? *Journal of Convention & Event Tourism* 14 (3), 217–235.
Lakoff, G. and Johnson, M. (2008) *Metaphors We Live By*. University of Chicago Press.
Nixon, R. (2011) *Slow Violence and the Environmentalism of the Poor*. Harvard University Press.
Norris, J. and Sawyer, R.D. (2016) Toward a dialogic methodology. In J. Norris, R.D. Sawyer and D. Lund (eds) *Duoethnography* (pp. 9–40). Routledge.
Paulus, T., Woodside, M. and Ziegler, M. (2008) Extending the conversation: Qualitative research as dialogic collaborative process. *Qualitative Report* 13 (2), 226–243.
Philo, C. and Wilbert, C. (2000) *Animal Spaces, Beastly Places*. Routledge.
Proctor, H. (2012) Animal sentience: Where are we and where are we heading? *Animals* 2 (4), 628–639.
Rathmayr, B. (1991) Jäger, Feldherrn, Vogelkundler, Detektive und Wanderer: Zur Metaphorik sozialwissenschaftlicher Forschung [Hunters, generals, ornithologists, detectivs and ramblers: Metaphors in social science]. In T. Hug (ed.) *Erziehungswissenschaft als Lebensform. Theoretische und erfahrungsreflexive Beiträge zur Hochschuldidaktik und Wissenschaftsforschung* (pp. 143–155). Österreichischer StudienVerlag.
Rickly, J.M. and Kline, C. (2021) *Exploring Non-Human Work in Tourism: From Beasts of Burden to Animal Ambassadors*. Walter de Gruyter GmbH & Co KG.
Sinha, S. and Back, L. (2014) Making methods sociable: Dialogue, ethics and authorship in qualitative research. *Qualitative Research* 14 (4), 473–487.
Schmitt, R. (2005) Systematic metaphor analysis as a method of qualitative research. *Qualitative Report* 10 (2), 358–394.
Sims, R. (2009) Food, place and authenticity: Local food and the sustainable tourism experience. *Journal of Sustainable Tourism* 17 (3), 321–336.
Stinson, M.J., Grimwood, B.S. and Caton, K. (2021) Becoming common plantain: Metaphor, settler responsibility, and decolonizing tourism. *Journal of Sustainable Tourism* 29 (2–3), 234–252.
Tomassini, L., Bertella, G., Grasso, C. and Lenzi, C. (2022) The space of animal justice in wildlife sanctuaries: A posthuman perspective. *Journal of Ecotourism* 23 (1), 38–55.
Tomassini, L. and Bertella, G. (2023) The human gaze at animals and the missing animal gaze in tourism studies. *Current Issues in Tourism* 27 (3), 350–355.
Tovey, H. (2003) Theorising nature and society in sociology: The invisibility of animals. *Sociologia Ruralis* 43 (3), 196–215.
Tully, P.A. and Carr, N. (2021) Farm animals' participation in tourism experiences: A time for proper respect. In J.M. Rickly and C. Kline (eds) *Exploring Non-Human Work in Tourism: From Beasts of Burden to Animal Ambassadors* (pp. 83–100). De Gruyter.
Venegas, G.J.Q. and López, Á.L. (2018) The (unethical) consumption of a newborn animal: Cabrito as a tourist and recreational dish in Monterrey, *Mexico*. In C. Kline (ed.) *Animals, Food, and Tourism* (pp. 36–51). Routledge.

Vickers, M.H. (2010) The creation of fiction to share other truths and different viewpoints: A creative journey and an interpretive pocess. *Qualitative Inquiry* 16 (7), 556–565.

Von Essen, E., Lindsjö, J. and Berg, C. (2020) Instagranimal: Animal welfare and animal ethics challenges of animal-based tourism. *Animals* 10 (10), 1830.

Winter, C. (2020) A review of research into animal ethics in tourism: Launching the annals of tourism research curated collection on animal ethics in tourism. *Annals of Tourism Research* 84, 102989.

Yudina, O. and Fennell, D.A. (2013) Ecofeminism in the tourism context: A discussion of the use of other-than-human animals as food in tourism. *Tourism Recreation Research* 38 (1), 55–69.

7 Human and More-Than-Human Animal Mobilities in an Insecure World: A Critique of (De) Humanising Narratives and Implications for Alternative Futures

Meghan L. Muldoon

Introduction

Mobility is at the center of all touristic practices and discourses. Unequal access to mobility, and the (lack of) freedom this confers, have been given increasing academic attention in many fields of late, including that of critical tourism studies (Bianchi, 2006; Hannam *et al.*, 2014). Tourism mobilities often go hand-in-hand with notions of leisure, recreation and fun. Increasingly, the water channels and beaches of Western and Southern Europe are witnessing a different type of arrivals, and ones that are portrayed as disruptive of touristic hedonism and the governments who rely on the associated profits. Expanding global political instabilities across many parts of Africa, the Middle East and Asia has led to mounting concerns over the conflicts that arise when the mass movements of peoples fleeing unsafe homelands arrive seeking safety on unwelcoming shores, and to much political and social debate on all sides of the political spectrum.

Perspectives on how best to support people fleeing violence and hardship are prevalent across academic and popular literatures, however the debate regarding more-than-human animal mobilities is less common (Blaer, 2022; Fennell, 2012). While the tourism literature has increasingly turned its attention to practices of utilizing nonhuman animals for touristic purposes (cf. Cohen, 2012; Kline, 2018), few, with the notable exception of Blaer (2022), have examined the link between tourism, volunteering and more-than-human animal mobility. For this research, the author interviewed residents of the Netherlands who have adopted dogs as animal companions from other countries perceived as having 'problem' dog populations, including Greece, Romania, Ukraine and Thailand. All of the participants actively sought to rescue dogs from shelters in countries where dogs are perceived as having a lowered quality of life and shortened life expectancies. Many of the participants chose to adopt a dog after having encountered stray dogs while on holidays in one of these countries or having visited an animal shelter while on vacation. Through the use of a critical discourse analysis, these discourses of more-than-human animal adoption and mobility are contrasted with discourses in the media and political debates regarding the sheltering and mobilities of humans faced with violence and limited life expectancies.

The purpose of this study is not to make a direct comparison between the mobilities of rescued and adopted dogs and refugee peoples: there are a multitude of structural, political, cultural and economic reasons why such a comparison would be nonsensical. Rather, I examine the ways in which language is deployed which results in the (subconscious? unconscious? intentional?) humanizing of rescued dogs and the dehumanizing of refugee humans in order to justify their (im)mobilities. As a result, this research aims to shed light on how human and more-than-human animal lives are (de)humanized in popular discourses and poses critical questions regarding future mobilities in tourism and beyond in Europe. In doing so, this chapter concludes by drawing implications for possible alternative futures based on typologies of future studies (Bergman *et al.*, 2010) and critical futures research (Inayatullah, 1998).

Tourism and Mobilities

The mobilities paradigm in tourism is conceptualized as encompassing both the local and everyday movements of people through time and space as well as the exoticized tourism as portrayed in our collective tourism imaginaries and experiences. As per Hannam *et al.* (2014), '[t]ourism is then analysed not as an ephemeral aspect of social life that is practiced outside normal, everyday life. Rather it is seen as integral to wider processes of economic and political development processes and even constitutive of everyday life' (2014: 172). Indeed, it is concerned with the larger geopolitical forces that constitute and are constituted of the worldmaking practices embedded within the local and international movements of humans, objects, information, capital and more-than-human animals (Bianchi, 2007; Hannam *et al.*, 2014; Hannam *et al.*, 2006; Hollinshead, 2009). Tourism mobilities are increasingly being understood as operating in synchronicity with broader transnational paradigms, including migration, climate change, and unequal access to resources and the necessities of life, and thereby demanding that they be considered simultaneously (Hannam *et al.*, 2014; King & Christou, 2000). For Bianchi (2007), tourism mobilities are inseparable from geopolitical realities at the destination and can lead to conflict whereby 'local grievances (linked to poverty, ethnicity, or questions of religious identity) and wider geopolitical imperatives "collide"' (Bianchi, 2007: 70).

In their collection, *Intersections of Tourism, Migration, and Exile*, Bloch and Adams (2022) strive to challenge the unproblematized siloing of these forms of human spatial mobility. Visiting friends and relatives (VFR), trips to established migrant and refugee communities, for example, Chinatowns, and movements towards spaces where refugees find themselves immobile and constrained often take the form of social justice, medical, or volunteer tourism (Adams & Bloch, 2022).

Mobility is not merely the uncontested movement between point *a* and point *b* across space and time, however. Since space is being continually (re)constructed as the product of social relations, in moving one is also contributing to '*alter* space, to participate in its continuous production… Space and place emerge through active material practices' (Massey, 2005: 117–8). Massey has argued that places ought to be understood as 'dynamic and open nodes that are in process, being cross-cut by a diverse array of productive movements which embody a complex geography of power relations' (as cited in Merriman, 2012: 2). Thus, multiple, and complex mobilities often constitute a 'fluid interdependence' (Sheller & Urry, 2006: 12).

The Refugees (in) Crisis

A growing number of global conflicts in places including Somalia, Gaza, Libya, South Sudan, Afghanistan and Syria, among many others, have led to a growing number of refugees seeking safety and stability at Europe's points of entry. Media reports and government spokespersons refer often to the 'crisis' triggered by increasing numbers of refugees. Many migrants are forced to pay exorbitant amounts of money to people smugglers for the chance to land on Europe's shores, travelling often in overcrowded and wildly unsafe boats and rafts. And many of their intended destinations, including Greece, Italy, Malta, Spain and the south of France, are regions heavily dependent on their respective tourism industries.

Research regarding conflict arising from tourist and refugee encounters has proliferated (cf. Cirer-Costa, 2017; Ivanov & Stavrinoudis, 2018; Pappas & Papatheodorou, 2017; Zenker *et al.*, 2019). And while many residents, including those on Lampedusa Island who were nominated for a Nobel Peace prize in 2016 for their reception of migrants (Marsi, 2023), have taken incredible measures to support the refugees, including knitting classes, maternity wards, and public art displaying support (Marsi, 2023), anti-migrant rhetoric has proliferated throughout western Europe as governments struggle to manage and pay for the influx of newcomers. Right-wing national leaders, including Georgia Meloni in Italy, Kyriakos Mitsotakis in Greece, Rishi Sunak in the UK and Geert Wilders in the Netherlands have capitalized on and promulgated anti-refugee sentiments. Efforts to restrict the arrival of refugees crossing from North Africa have included the EU funding the Libyan coast guard to reduce the number of boats arriving in Europe, and whom have been witnessed ramming boats overloaded with refugees (Al Jazeera, 2023), European navies and coast guards actively ignoring refugee ships in distress (Beake, 2023; El Idrissi & Pascual, 2023), and turning away boats as they arrive in European waters (Trilling, 2023), an act in contravention of the 1951 UN Convention Relating to the Status of Refugees.

Measures to disincentivize refugees' arrivals in the UK have included a proposal to relocate migrants – regardless of their nation of origin – to Rwanda (BBC, 2023) and housing up to 500 migrant men on a former prison barge in Dorset (Couzens *et al.*, 2023). As many European nations resist pressure to accept more migrants within their borders, refugee men, women, and children become increasingly immobile in makeshift refugee camps, relocation centers, and even cages in the Tripoli Zoo (Fadel, 2013).

Companion Non-Human Animal Rescue Tourism

Non-human animal tourism has received a plethora of academic attention in recent years (cf. Fennell, 2012, 2022a, 2022b, 2023, 2024; Fennell *et al.*, 2023; Fennell & Guo, 2023; Fennell & Thompson, 2021; Hoarau-Heemstra & Kline, 2022; Kline, 2018a, 2018b, 2021; Kline & Fischer, 2023; Taylor *et al.*, 2018), including – of course – chapters shared in this volume. Many focus on wildlife viewing of exoticized animals, including best practices for optimal viewing (Hammit *et al.*, 1993; Higham *et al.*, 2008; LeBoeuf & Campagna, 2013; Knight, 2009), the economic benefits (Mitchell, 1968) and emotional affectation (McIntosh & Wright, 2017). Other studies are concerned with domesticated animals used as laborers in the tourism industry; this includes, for a very few examples, elephant rides in Thailand (Taylor *et al.*, 2018), sled-dog attractions in Canada (Fennell, 2022a), and tigers held in captivity and sedated in order to 'pose' for photographs with tourists (Cohen, 2012). As household pets are becoming increasingly more important members of many families in the 21st century, many more recent studies have turned to tourists being accompanied by their pets to their travel destinations (Carr, 2017; Hidalgo-Fernández *et al.*, 2023; Tomassini, 2022), tourism destinations taking steps to become more 'pet-friendly', (Alves *et al.*, 2022; Mlakar & Korže, 2022), and even pet medical tourism (Wu & Chang, 2023).

With one notable exception, very few studies have focused on more-than-human animal mobilities in the context of rescue and adoption.

Animal rescue organizations have increasingly turned to attracting volunteer tourists to help with the day-to-day operations at shelters providing food, housing, medical care and affection for unhoused domesticated animals, often in the absence of any financial support outside of donations. Often these are operated by altruistic individuals who recognize the plight of urban animals and the problems caused by unchecked animal populations. These start off small, but as they become more well-known, local people increasingly begin to drop off animals that are unwanted, ill, have been struck by cars, have been abused, and so on. Some of the larger animal shelters, including Soi Dog in Thailand and Takis in Greece have millions of followers on social media (looking only at Facebook, Takis has 950,000 followers and Soi Dog has 1.3 million as of January 2024). Tourists may choose to spend one day of their holidays volunteering at the shelters. Others will make the volunteering experience the focus of their travel and remain with the shelter for weeks or months.

In her research, Blaer (2023) looked at the case of Animal Aid Unlimited (AAU). AAU, located in a small village in Udaipur, was founded by Americans Erika and Jim Abrams, who moved to India after travelling there and founded AAU in 2002 after becoming aware of the poor health and limited life outcomes of unhoused domesticated animals. Their mission is guided by '"five freedoms": the freedom from (1) hunger or thirst, (2) discomfort, (3) pain, injury, or disease, (4) fear and distress..., and (5) freedom to express (most) normal behaviour by providing sufficient space, proper facilities, and company of the animal's own kind' (Blaer, 2023: 6). AAU welcomes volunteers, who provide care and affection for the animals, however the organization, being first and foremost animal-centered, does not rely on volunteers for their operations. As such, they do not detract from their mission by providing food, lodging or transportation for the volunteers; those who wish to support the work being done with AAU must make their own arrangements. In return, the staff at AAU can focus on providing the medical and day-to-day necessities for the animals, and the awareness of their organization's work is spread through the social media activities of the tourists (Blaer, 2023). Through animal volunteer tourism, Blaer highlights the ways in which these rescue operations 'can offer a counter experience to the largely disturbing discourse of the treatment of many animals in tourism' (2023: 2).

All three of the more-than-human animal rescue organizations mentioned above, and many more, use their websites and social media postings to showcase dogs that are available for adoption. People choose to adopt dogs from rescue organizations for a myriad of reasons, many of which are explored below. In the following sections, I describe research undertaken in two parts. First, unstructured interviews were conducted with eight men and women living in the Netherlands who had adopted rescued dogs from other countries, including Romania, Greece, Ukraine and Thailand. In the second part of the study, news articles from the year

2023 relating to migrants and refugees arriving in Europe were collected and analyzed using a critical discourse analysis. In the following section, I describe the methods I used to undertake this study, followed by a presentation and discussion of the findings.

Methods

Data for both components of the study were collected simultaneously over a period spanning from September 2023 to January 2024. For the interview data, a mix of purposive and snowball sampling was used. Two interviewees are people known to the researcher from the local dog park and a third is a colleague of a friend of the researcher. From these three interviewees, five other adoptee dog companions were referred and agreed to participate in the study. No constraints were placed on where the dogs were adopted from, however they had to have originated outside of the Netherlands, where the participants reside. Seven of the participants were women and one was a man. In addition, two participants have worked as volunteers in their local communities, helping to identify and evaluate suitable homes for dogs who are being adopted from other countries, two have acted as fosters to a number of dogs when they first arrived in the Netherlands, and a fifth is a volunteer with the Dutch animals' political party (Partij voor de dieren).[1] Two interviews were conducted while walking in the dog park, a joint interview was held in one participant's apartment, and the remainder took place in local restaurants. All interviews were conducted in English, the native language of the researcher and a very accessible second language to most of the participants. Two of the participants who felt less confident in English chose to be interviewed together, so they could support one another. The interviews lasted between one and two hours and were undertaken in an unstructured format in order to allow for conversational dialogue (Chauhan, 2022). All interviews were audio-recorded and transcribed. Data analysis was undertaken using line-by-line and descriptive coding in order to identify the themes described below.

Data from the media outlets (via X, formerly known as Twitter) were analyzed using a critical discourse analysis (CDA). Discourse analysis 'examines how discourses are constituted and circulated within texts and representation, which in turn function to produce a particular understanding or knowledge about the world that is accepted as "truth"' (Waitt, 2005: 168). With CDA, texts are critically examined for effects of truth, inconsistencies, and exclusions or silences, because it is important 'to look beyond what was contained within the texts and to consider what was left out or not obvious, as well as the mechanisms that enable these exclusions' (Grimwood *et al.*, 2015: 28; Hannam & Knox, 2005; Rose, 2007; Waitt, 2005).

The data for the critical discourse analysis of media related to refugees and migrants to Europe were collected from the social media site X. X was chosen as the source of the data because media outlets use the site to disseminate news articles, many people refer to X as their first source for current events, and X includes an advanced search function which allows for multiple parameters and saves having to search multiple news media websites or databases. The key search terms used were 'migrant' and 'refugee'. A number of exclusion terms were also applied to the search, in an effort to restrict the search to news stories relating to Europe. These were: Israel, Israeli, Palestine, Palestinian, Gaza, West Bank, Ukraine, USA, Mexico, India and Rohingya. In an effort to include English-language news articles from multiple countries that have been entry points for migrants and refugees arriving from (primarily) North Africa, the search focused on articles from the X accounts of The BBC World, The Guardian, Le Monde, Euro Weekly News, Al Jazeera, The Portugal News, The Times of Malta, Total Croatia and The Local Italy.

These online news sites were selected as I wished to include at least one English language news source from each of the major refugee receiving countries in Europe. These range from long-established highly credentialled and well-established national and/or transnational news sources (e.g. The BBC, Al Jazeera) to more ad hoc local news sources, which were identified as the only accessible news sites in that location (e.g. The Times of Malta, The Local Italy). Finally, the date range for the search included articles published between January 1, 2023 and November 9, 2023, in order to make the volume of articles manageable. In total, 147 articles were included in the study, with the majority being sourced from The BBC, The Guardian, and, to a lesser extent, Al Jazeera.

Findings

The findings of the study are presented in two sections, representing the two different phases of the study; the interviews with adoptive companions of dogs from outside the Netherlands, and the critical discourse analysis of the news texts posted on X. A discussion of the significance of these findings in relation to mobilities and tourism futures follows.

Companion nonhuman animal rescue tourism

Through an analysis of the interview data, a number of themes emerged. These have been categorized as: *For the Love of Dogs*, *Patience*, *Logistics* and *Travel and Nonhuman Animal Companions*. Each will be presented in detail below.

For the love of dogs

All of the participants identified themselves as dog lovers. Many grew up with dogs and considered them to be a part of the family. Five of the eight participants had their dogs in their social media profile pictures, which was apparent as Facebook was the primary means of communication in the recruitment of participants. A number of participants explained that there are cultural differences in that in the Netherlands dogs are a part of the family, but it is not so in the countries where many of the dogs come from. As one participant explained:

> *For them it is not normal to have a dog in their home and pet them like we do. I have dogs with holes in their ears. Those are dogs coming from shelters and they're on the list to kill. Sometimes we have dogs with burn holes on their noses. Dogs with tails that were cut off.*

Another described what they thought was the reason for their dog's trauma and resulting fear and antisocial behaviour:

> *I think they hunted her and tried to find her with flashlights. She is absolutely afraid of any lights. And I think she had a real beating.*

Two participants described adopting dogs that were at risk of being euthanized. One had been abandoned at the local shelter and the shelter staff was unable to care for it because of the dog's behavioral issues. Another was adopted from Ukraine, after a shelter there reached out and shared that they were no longer able to care for the dogs following the Russian invasion. This participant, in particular, had two previous negative experiences with adopting dogs from Romania, one who was so traumatized it was unable to live in an urban environment and a second who died following veterinary surgery less than six months after having been adopted. Nevertheless, this participant was still willing to open their home to a rescue dog less than three months after their previous dog passed away because it was in need of a home. One participant fell in love with a dog while volunteering at an elephant sanctuary while in Thailand, and undertook the arduous process of bringing the dog back to the Netherlands. As they described:

> *If you really love a dog [the cost] doesn't matter. Because he was also disabled and we needed to do a lot of CT [Computed Tomography] scans and all those thing on him. So it was a very expensive dog anyway. I spent so much money on him, but I didn't care because I really loved this guy. And it's just your best friend.*

Mostly, the participants in this research were motivated by a desire to give a dog a second – or often first – chance at a safe, stable home and a loving family. The participant who adopted the dog who had been abandoned at a local shelter, described him in this way:

> *He was horrible. But he was happy.*

Another spoke of the drive to adopt:

> *We were looking to find an adoption dog. Because we want to give a dog a good home. We of course want a dog, but we want to have a dog that has no future. That was our goal.*

They went on to add:

> *We wanted to have one of the dogs, at least one of the dogs from there who have a hellish life to have a good life. And we're there for the dog. She's not there for us.*

Patience

All the participants described their rescue dogs as having come to them with varying degrees of trauma. Immense patience and understanding are required in order that the dogs may come to feel safe and secure in their new homes. And some never do. As mentioned above, one participant had no choice but to rehome their dog when she was unable to cope with life in the city. Another made the decision to return a foster to Greece – a difficult decision to make, they said, but ultimately, they felt he would be happier on a mountainside in Greece (where the shelter is located) than in a city in northern Europe. Participants describe the dogs as being highly fearful, and having extreme reactions to fireworks, other dogs, elevators, unfamiliar people, children, even cars. One participant felt strongly that first-time pet owners or families that want a dog to love them should not consider adopting from abroad:

> *You shouldn't start with a foreign dog, especially one from Spain or Romania. Okay, because you don't have an instant pet. Normally, if you get a puppy, you have an instant pet. They're not. They're skittish, they don't want anything to do with you. And people don't understand it.*

This participant explained that it has taken their dog nearly two years to relax and feel safe in their family, to the point that she will occasionally seek out human contact. But when unfamiliar people are in the home, her instinct is to hide under the couch:

> *We say she's always on, she's never relaxed. It's getting better, a bit, she's two now, but still very nervous. Around us she's perfectly fine, but around other people, strange people, she's under the couch. And we let her. I mean, if that's her way of coping with it. And people talk to us about you have to do something about it, no, I have a dog for me and for the dog, not for you. And if that's where the dog needs to be to feel safe, yeah, no worries.*

One participant described the process that they undertook in an effort to build up the trust with their adopted dog from Greece to the point that she would allow them to touch her:

> *I couldn't get a collar on her. If I came anywhere near her she would yell, like I was beating her...What I did, I started making meat soup and mixing it in with her food, plus she was real skinny, you know? And then I would get her collar and she'd have to try to get the food without letting the collar touch her...So I just sat with her and then occasionally, once she started getting used to it, then she'd have to put her head on the collar to eat. And you know, it was not days, weeks, you know? ...I have it on Facebook when the first time she let me touch her.*

For a number of participants, the COVID-19 pandemic was an opportunity to adopt a dog, because they were working from home and many rescue organizations will not house a dog with a family that is out of the home full-time, plus all the extra time and effort it takes to acclimatize the dog to living in a home with people. Some have never been leash trained or had a collar on. Some animal rescue organizations recommend a double harness, because many rescue dogs have started at unexpected noises or other dogs, slipped their collars, and ran away, where there is a risk of them falling into a canal or being struck by a car. One participant described how they book a hotel room in another community that does not permit fireworks on New Year's Eve every year. The fireworks are quite traumatic for many of the adopted dogs, with one participant describing how they ensure that their dog has a thunder shirt[2] for New Year's Eve, and another has taken to giving their dog canine CBD oil, upon the recommendation of another dog's human companion.

Logistics

Rescue organizations with longstanding reputations have extensive criteria for rehoming shelter dogs from abroad. They want to know how many hours a future family works away from home, whether or not the home has a garden, whether there are any other dogs in the family, whether there are children, and what ages they are, among other criteria. The evaluation often also includes a home visit, and one participant described visiting potential future families with their own dog to see how the family's existing dog interacted with another dog. Another volunteer explained that they would be in contact with the family once a week for the first month, then once a month for the next three months. Many participants revealed that the animal shelter in their community does not accept dogs who have been adopted from abroad, so it is critical that the new home is a good fit.

Other participants have described less thorough adoption processes that have resulted in less positive outcomes. Some were not informed about the need for double harnesses or were not given true information about the dog's disposition due to their trauma. One participant, who fell in love with their dog at the elephant sanctuary in Thailand, described how it was incredibly challenging navigating the immigration process in both countries, something that the elephant sanctuary had no previous experience with. Some rescue centres, particularly those operating as a passion project as opposed to a more formal organization, have sent over dogs that were not suitable to be rehomed, and for this reason one participant no longer volunteers with them.

Travel and nonhuman animal companions

Physically transporting rescued companions from abroad can also pose logistical challenges. One rescue centre in Romania has a van which travels to the Netherlands with several dogs each week. I was also told of a future family that drove from Switzerland to adopt their new companion who had arrived in the Netherlands from Greece. For some volunteers, an important part of their role is identifying potential travel companions for dogs on their flights. The adopting families will pay the cost of the tourist's return flight, and in return, the tourist will ensure the safe travel of the dog(s) being adopted. For individuals adopting from distant places such as Thailand, this significantly adds to the financial burden of adopting a dog.

While the majority of participants in this study made the decision to adopt from abroad and were able to connect with their future companion via an internet search, many described having seen unhoused animals on the streets while they were travelling. One participant shared that they vacationed in Greece and was horrified when they saw the living conditions of the unhoused dogs at the destination. Upon returning home to the Netherlands, they and their partner immediately began the processes of adopting a dog from Greece.

Another, as mentioned above, met their dog while a volunteer tourist at an elephant sanctuary, a tourism practice that they engaged in for several months a year for four or five years. Twice, they travelled to Thailand to volunteer and returned to the Netherlands with dogs for adoption. They also connected with their network of other volunteers from around the world and encouraged them to adopt. Although the sanctuary had not been developed with dogs in mind, many dogs are rescued from the illegal dog meat trade and the sanctuary has become a resource for these animals, now housing hundreds of dogs. Another participant shared that they travelled to Greece to volunteer at a dog sanctuary there, but they did not enjoy the environment at the shelter and did not remain for long.

Critical discourse analysis of refugee and migrant news stories

Shifting now to the findings of the critical discourse analysis of media stories, several observations regarding the deployment of language in these stories were identified. Of particular interest to this story, were the way in which The BBC, in particular, labelled the arrival of migrants in Europe as a 'crisis', the political orientations of the media sites included in this study, and how that is perceived to have oriented their reporting, as well as the public comments that accompanied the articles' postings on X.

The 'migrant crisis'

The majority of the articles considered in this research were filed under certain categories. For example, The Guardian articles that were considered were filed under 'migration' or 'immigration and asylum'. In Le Monde, most articles were filed under either 'international' or the name of the country featured in the article, which was the same for The Euro Weekly News. Al Jazeera used the categories 'refugees' and 'migrants'. The Local Italy, from which only one article was collected, used the category 'migrant crisis'. In the same vein, The BBC, which accounted for 47% of the articles collected, classified them all under the heading 'European migrant crisis'. Framing the articles in such a way, before the reader has had an opportunity to engage with the content of the article, already establishes in the reader's mind that the presence of migrants in Europe is problematic, a menace threatening Europe. This choice of phrasing presages any decision for the reader about whether or not the arrival of migrants in Europe may have a positive effect, or whether the situations that people are fleeing are so dire that they have little to no other choice. Instead, categorizing migrants' arrivals in Europe as a 'crisis' leaves little space for European readers to consider them as anything but a problem that requires solving.

Newspaper orientations

Immediately apparent upon engaging with a number of articles was that the journalists' perspectives largely mirrored my own regarding the vilification and efforts at turning back migrants seen from European politicians, coast guards, and navies. A choice was made in this data collection to look to newspapers from a number of migrant-receiving countries. In so doing, I selected media sites that I was most familiar with, meaning those that I turn to most often to receive my own news.

The articles collected through this research were highly critical of the profits garnered by private hotels in Britain which house refugee claimants, the decision to house migrants on a former prison barge, the ramming of a migrant boat by the Libyan coast guard, the failure of coast guards and navies to respond to migrant ships in distress, leading to the deaths of hundreds of people, and the racist rhetoric used in defence of governments' anti-migration efforts from politicians such as Suella Braverman, the then-Home Secretary to the British government. These news stories speak of tragedy and the loss of human life, as well as the dehumanization of refugees housed in overcrowded refugee camps, prevented by guards from speaking with reporters or being relocated to a processing centre in Albania, being physically assaulted and pushed back at border crossings, or held in a former zoo.

The articles also pointed to many politicians speaking about how the majority of refugee arrivals in Europe are men, implying questions as to the refugees' intentions and possibly posing a threat to European safety and security. The articles point to the danger implicit in such rhetoric, while at once amplifying such speech, potentially contributing to anti-immigrant movements such as those seen in Berlin and across Germany on New Year's Eve 2022.

As the majority of the articles considered in this study emphasized the lack of human rights afforded to human migrants and the repudiation of the UN Convention for Refugees by a number of governments, including those of Greece, Spain, Italy and Britain, future research may be interested in focusing on media sources from all sides of the political spectrum in order to gain a more diverse perspective on media reporting regarding human migration to Europe.

Public comments

One of the features of the social media site X is that it allows for public commenting on any post. Despite the articles selected for this study largely being critical of the violence and dehumanization migrants and refugees face in attempting to find a new life in Europe, the majority of the public comments on these articles express a public sentiment strongly against the arrivals of newcomers from the Global South. For instance, in one positing, headlined 'Negligence may have led to 36 migrant deaths in Canary Islands, lawsuit alleges' shared by The BBC on July 12, 2023, only one tweet (of those not blocked due to offensive content) expressed support for the human lives lost:

> *All the inhuman, racist monsters attacking the victims should be taken from their comfortable existence of spreading hate behind computer screens and put afloat on a dinghy in the Atlantic.*

The other comments on the article are, in order:

> *Negligence of the migrants?*
> *The Canarian authorities will struggle with the "we thought Morocco were sending help" but essentially this might serve as a warning of the dangers of this.*
> *Surely that's misadventure.*
> *Negligence is on them, nobody makes them come the way they do..if they have papers then come in the right way…*

Negligence by whom?
We should do the same in the Channel
Yes, the migrants were negligent with their own lives by getting on that boat
They should've stayed home
Negligence by European authorities may have cost them Europe.
Negligence (3 x cry-laughing emoji) stop using the blame card...nothing to do with Spain...for a start they new the risks...they should of stayed at home...We need this policy in the channel....(4 x St. George's cross flag emojis, 2 x finger and thumb circle emojis)
Negligence? Like in stupidity? Henlon frowns!
You've done documentary on the peaple smugglers and the illegal immigrants who both acknowledge the trip is dangerous at that point stop blaming society and tell them to blame themselves (union jack flag emoji, eyebrow raised emoji)
36? Saved alot of money
Why the BBC falsely labels illegal immigrants as 'migrants'?
I knew someone would find a way to blame Whitey.

The majority of the comments on the articles are mocking of the deaths of migrants attempting to cross in boats, supportive of government policies to turn back or deport refugee claimants, and, most significantly, pointedly directed at the 'choice' people have made to leave their home countries, that they know the risks, and that they are responsible for the harms that befall them for having chosen to leave home. Racism also plays a prominent role in many of the comments. A small selection of the more abusive and dehumanizing of the comments includes:

The total white population on earth is about 700 million or 8.5% but the 7 billion non-whites who hates us so much are invading the white western world and doing they're damndest to live with us...Get the picture I don't
All countries should resign from the UN convention on refugees. They need to stop all those middle eastern and African country shoppers invading everywhere
200 rapists less in Europe
If they stayed home, they could be in the pub
You'd think one of them would know how to steer a boat it's not hard
Calarification: after an island of 6,000 is invaded by 10,000 males

Discussion

To be very clear, the purpose of this chapter is in no way to be critical of people who love dogs and undertake supreme efforts to bring them into their homes and make them feel safe and loved. All dogs need love and companionship and their abuse and neglect are appalling. I have nothing but admiration and respect for the participants in this study who spoke of their love for their more-than-human companions. Rather, this study aims to shine a light on how language is deployed in order to explain why one would make large personal and financial sacrifices to free a dog from terrible living conditions and then is also used to dehumanize and mock people fleeing whatever hardships and violence they were subjected to in their home countries.

In their volume, *Like an Animal: Critical Animal Studies approaches to borders, displacement, and othering*, Khazaal and Almiròn (2021) discuss the ways in which asylum seekers and refugees are othered through discourses which dehumanize them and justify their exclusion in order to preserve and protect the homogeneity of the host society. Through this othering, the host society is constructed as racially and/or morally superior to that of the migrants' home societies. The field of Critical Animal Studies (CAS) aims to critically examine the ethics of human relationships with other-than-human animals from a political and intersectional perspective. The present study builds upon the work elaborated by Khazaal and Almiròn and their contributors in bringing together 'the fields of dehumanization (and critical refugee studies), critical border studies, and critical animal studies…to think about how speciesism plays a formative role in questions of refuge, borders, displacement, migration, and dehumanization…to open a new space to explore *speciesism-as framework*' (Khazaal, 2021: 5–6).

Khazaal and Almiròn (2021) aim to blur the hard speciesist divide separating human and nonhuman animals, characterizing the 'soft divide' between the two as dehumanized out-groups, including refugees (Khazaal, 2021: 7). As Khazaal writes, '[Critical Animal Studies] perspectives acknowledge the social pathology of dehumanizing human refugees as "animals" inherent in ethnocentrism and its consequences – denial of dignity and autonomy, unequal treatment, and increased control' (2021: 7). Benesch (2018) refers to the spreading of language that convinces people to commit violence against vulnerable others as 'dangerous speech.' Khazaal refers to dangerous speech as 'part of a millennia-old practice of racism and dehumanization across all human societies' (2021: 9). She refers to the example of increased persecution of Jews in Nazi Germany in areas which had access to better radio reception, and therefore increased access to the dangerous speech spread by Nazi rhetoric. Dangerous speech leads to harmful ideology, wherein a dominant group is permitted (encouraged?) to subjugate marginalized or vulnerable groups in order to justify the unjust social order (Khazaal, 2021). Harmful ideologies create an atmosphere in which speciesism, as a form of Othering, is allowed to perpetuate and flourish. Speciesism signals that belonging to a particular species has a moral value and that being a member of the Other species is repugnant. According to Khazaal (2021), 'speciesism is used to justify and perpetuate catastrophic forms of oppression against nonhuman animals' (2021: 6). Speciesism underlies other forms of oppression and violence, including, in this case, anti-migrant and refugee racism. Of particular significance to this study are the ways in which dehumanizing language is used to animalize and further justify the violence and expulsion of migrants. Recent examples include dehumanizing Rohingya Muslim refugees as '"African carp," "snakes," "mad dogs," "beasts," "wolves," "jackals", "human fleas," and "maggots"' (Khazaal, 2021: 9). Such dehumanizing language dates back at least as far back as to the Ancient Greeks, who referred to their slaves as 'beasts' (Khazaal, 2021).

According to Smith (2012), cognitive biases lead us to construct social hierarchies that separate human groups from one another, emboldened by political language that constructs the Other as positing a moral threat to society. The Other poses a threat not only to the moral fabric of the society but also an economic threat. As such, dehumanization 'isn't a blind spot of human evolution but a solution to the problems of guilt from killing and shame from violating one's own moral principles, i.e., hypocrisy' (Khazaal, 2021: 14–15; Smith, 2021). In order to overcome the 'conflict between fellow feeling and the love of money' political rhetoric is deployed via dangerous speech in order to reduce the members of the outgroup, that is, the constructed threat, as 'mere animals…to be skinned, castrated, and burned on the barbeque' (Smith, 2021: 107, 174; Khazaal, 2021). As demonstrated in this study, this dangerous speech (re)perpetuates the Othering of migrants and refugees to Europe, and casts them as violent predators looking to conquer and destroy the European way of life. The fact that much of the rhetoric, in particular coming from the commenting public, but also as reported in the politicians' speech, is undisguisedly racist against non-white refugees demonstrates just how violent and polarizing these discourses have become. And, tragically, discursive violence begets real-world violence, in the forms of the imprisoning of migrant bodies, forced relocations, massive, unacceptable losses of life at sea, starvation, risk of physical harm in the camps and violent assault at nearly every turn.

Conclusions: Dystopian and Utopian Scenarios and Alternative Futures

Humans have an infinite capacity for love and compassion, as demonstrated by the participants in this research, who opened their hearts and homes to dogs who had been starved, beaten, smuggled, burned and neglected. Their words describe the love that exists between people and their more-than-human companions, individualizing the plight of animals that have never known safety or security. In stark contrast, the language deployed by right-wing politicians in Europe – echoed by public discourse – dehumanizes people seeking safety, justifying their forced immobilities and the violence perpetrated against them.

Using Bergman *et al.*'s (2010) ontological typologies of futures studies, this chapter envisions two opposing scenarios regarding future human and nonhuman mobilities: a dystopian future, where exclusionary practices intensify, and a utopian future, where alternative approaches challenge dehumanization and speciesism. Additionally, applying Inayatullah's (1998) approach to critical futures research enables a deeper examination of the underlying worldviews and discourses that sustain these opposing realities.

Dystopian scenario: Dehumanising the other

In a dystopian trajectory, the language of dehumanization becomes further entrenched, reinforcing policies that justify the containment and exclusion of migrants. Dangerous speech continues to proliferate, shaping the perceptions of refugees as threats to economic stability and social cohesion. Echoing the ontological claims outlined by Bergman *et al.* (2010), this future operates within a deterministic framework, where existing inequalities are naturalized and self-perpetuating. Meanwhile, the trend of humanizing certain nonhuman animals persists – but in ways that serve anthropocentric and neoliberal agendas. The commodification of rescued animals through social media and tourism industries leads to an expansion of voluntourism initiatives, where privileged tourists 'save' animals while turning a blind eye to human suffering. This selective compassion reinforces the moral hierarchy that places nonhuman animals above displaced human populations. The critical questions raised in this study – about who is granted mobility and who remains immobilized – become starker as public empathy continues to fragment along speciesist and racial lines.

Utopian scenario: An inclusive understanding

Alternatively, an utopian future emerges from the principles of critical futures research (Inayatullah, 1998), which seeks to challenge deep-seated assumptions and create transformative possibilities. Here, speciesist and nationalist ideologies are critically interrogated, leading to more inclusive policies that recognize the intrinsic worth of all beings. Instead of weaponizing language to dehumanize refugees and migrants, discourses shift toward narratives of solidarity, mutual care and shared vulnerability. In this scenario, policies supporting ethical mobilities emerge. Governments, recognizing their moral and legal responsibilities, dismantle inhumane border policies and implement equitable asylum and migration systems. Public discourse moves beyond fear-based narratives, influenced by media frameworks that centre dignity and justice rather than exclusion and crisis.

Tourism, rather than reinforcing hierarchies of value, becomes a catalyst for cross-species and cross-cultural understanding. The adoption of critical futures research approaches (Inayatullah, 1998) in policymaking and academia helps unpack and challenge the dominant myths that sustain speciesism and xenophobia. Instead of reinforcing neoliberal frameworks that commodify animal welfare and human suffering, tourism practitioners and scholars actively participate in shaping ethical mobilities where care, rather than control, becomes the guiding principle.

Toward alternative futures

The competing visions outlined here underscore the urgency of addressing the ways in which language shapes mobility and exclusion. If the dystopian future is one in which immobilization and violence persist unchecked, then the utopian alternative requires intentional, systemic interventions to disrupt harmful ideologies. For scholars and policymakers alike, engaging with critical futures research offers a means to envision and enact change. The findings of this study reveal the need for a fundamental shift in how both human and nonhuman mobilities are framed within tourism and beyond. By integrating Bergman *et al.*'s (2010) typologies of futures and Inayatullah's (1998) critical futures research, this chapter calls for a deeper interrogation of dominant narratives and a commitment to futures where no being – human or nonhuman – is reduced to an object of exclusion or commodification. If we can make space in our hearts for traumatized and abandoned dogs – many of whom require patience, rehabilitation and care – then we must also find ways to extend that same compassion toward those seeking refuge. Perhaps, in reimagining tourism as a site of ethical engagement, tourists and scholars alike can become key actors in dismantling the structures of speciesism and xenophobia that shape the present, paving the way for more just and inclusive futures.

Notes

(1) https://www.partijvoordedieren.nl/ Their political objectives are: (1) Respect for animals; (2) More nature; (3) Food revolution; (4) Climate action!; (5) Health living environment; and (6) A different economy.

(2) A compression shirt that applies continuous, gentle pressure to soothe dogs and cats with stress and anxiety.

References

Al Jazeera (2023, Sept. 30) Rescue group decries Libya coastguard ramming into migrant boat. Available from https://www.aljazeera.com/news/2023/9/30/rescue-group-decries-libya-coast-guard-ramming-into-dinghy?sf182298297=1 (accessed January 13, 2025).

Alves, H.M., Sousa, B., Carvalho, A., Santos, V., Lopez Dias, Á. and Valeri, M. (2022) Encouraging brand attachment on consumer behaviour: Pet-friendly tourism segment. *Journal of Tourism, Heritage, and Services Marketing* 14 (8), 16–24.

Beake, N. (2023, 23 June) Greece ignored offer to monitor migrant boat, says EU border agency. *The BBC*. Available from https://www.bbc.com/news/world-europe-66003244?xtor=AL-72-%5Bpartner%5D-%5Bbbc.news.twitter%5D-%5Bheadline%5D-%5Bnews%5D-%5Bbizdev%5D-%5Bisapi%5D&at_link_type=web_link&at_link_

id=7BA588CE-11F4-11EE-B378-0FB3D772BE90&at_medium=social&at_format=link&at_campaign_type=owned&at_bbc_team=editorial&at_link_origin=BBCWorld&at_ptr_name=twitter&at_campaign=Social_Flow (accessed January 13, 2025).

Benesch, S. (2018) *Dangerous Speech: A Practical Guide. The Dangerous Speech Project.* Available at https://dangerousspeech.org/guide/ (accessed January 13, 2025).

Bergman, A., Karlsson, J.C. and Axelsson, J. (2010) Truths claims and explanatory claims – an ontological typology of futures studies. *Futures* 42 (8), 857–865.

Bianchi, R. (2006) Tourism and the globalization of fear: Analyzing the politics of risk and (in)security in global travel. *Tourism and Hospitality Research* 7 (1), 64–74.

Blaer, M. (2022) Animal rescue tourism: Digital technology-enhanced approaches to support voluntourist engagement, animal welfare, and rights. *Tourism Recreation Research* 49 (3), 471–485. https://10.1080/02508281.2022.2046991

Carr, N. (2017) Recognizing the position of the pet dog in tourism. *Annals of Tourism Research* 62, 112–113.

Chauhan, R. (2022) Unstructured interviews: Are they really all that bad? *Human Resources Development International* 25 (4), 474–487. https://doi.org/10.1080/13678868.2019.1603019

Cirer-Costa, J.C. (2017) Turbulence in Mediterranean tourism. *Tourism Management Perspectives* 22, 27–33.

Cohen, E. (2012) Tiger tourism: From shooting to petting. *Tourism Recreation Research* 37 (3), 193–204.

El Idrissi, A. and Pascual, J. (2023, Oct. 12) How the French navy tried to interfere in the 2012 Channel migrant shipwreck. *Le Monde*. https://www.lemonde.fr/en/les-decodeurs/article/2023/10/12/how-the-french-navy-tried-to-interfere-in-the-2021-channel-migrant-shipwreck_6168083_8.html?utm_term=Autofeed&utm_medium=Social&utm_source=Twitter#Echobox=1697133162 (accessed January 13, 2025).

Fadel, L. (2013, Nov. 2013) Tripoli zoo sees different kind of cage – one with migrants. *NPR*. https://www.npr.org/sections/parallels/2013/11/12/244550320/tripoli-zoo-sees-different-kind-of-cage-one-with-migrants (accessed January, 13, 2025).

Fennell, D. (2012) Tourism and animal rights. *Tourism Recreation Research* 37 (2), 157–166

Fennell, D. (2022a) Animal-informed consent: Sled-dog tourism as asymmetrical agential events. *Tourism Management* 93, 104584.

Fennell, D. (2022b) An animal welfare literacy framework for tourism. *Annals of Tourism Research* 96, 103461.

Fennell, D. (2024) *Tourism and Animal Ethics* (2nd edn). Routledge.

Fennell, D. and Thomsen, B. (2021) Tourism and animal suffering: Mapping the future. *Annals of Tourism Research* 91, 103317.

Fennell, D. and Guo, Y. (2023) Ubiquitous love or not. Animal welfare and animal informed consent in giant panda tourism. *Animals* 13 (4) , 718. https://doi.org/10.3390/ani13040718

Fennell, D., Coose, S. and Moorhouse, T.P. (2023) An animal welfare syllabus for welfare tourism. *Journal of Sustainable Tourism* 31 (5), 1071–1089.

Grimwood, B.S.R., Yudina, O., Muldoon, M. and Qiu, J. (2015) Responsibility in tourism: A discursive analysis. *Annals of Tourism Research* 50, 22–38.

Hammit, W.E., Dulin, J.N. and Wells, G.R. (1993) Determinants of quality wildlife viewing in Great Smoky Mountains National Park. *Wildlife Society Bulletin (1973–2006)*, 21 (1), 21–30. http://www.jstor.org/stable/3783356

Hannam, K. and Knox, D. (2005) Discourse analysis in tourism research: A critical perspective. *Tourism Recreation Research* 30 (2), 23–30.

Hannam, K., Sheller, M. and Urry, J. (2006) Mobilities, immobilities, and moorings. *Mobilities* 1 (1), 1–22. https://doi.org/10.1080/17450100500489189

Hannam, K., Butler, G. and Paris, C.M. (2014) Developments and key issues in tourism mobilities. *Annals of Tourism Research* 44, 171–185.

Hidalgo-Fernández, Á., Moral-Cuadra, S., Menor-Campos, A. and Lopez-Guzman, T. (2023) Pet tourism: Motivations and assessment in the destinations. *Consumer Behaviour in Tourism and Hospitality* 18 (3), 335–350.
Higham, J.E.S., Lussea, D. and Hendry, W. (2008) Wildlife viewing: The significance of viewing platforms. *Journal of Ecotourism* 7 (2–3), 137–146.
Hollinshead, K. (2009) The 'worldmaking' prodigy of tourism: The reach of power of tourism in the dynamics of change and transformation. *Tourism Analysis* 14 (1), 139–152.
Hoarau-Heemstra, H. and Kline, C. (2022) Making kin and making sense of human-animal relations in tourism. *Ecological Economics* 196, 107396.
Inayatullah, S. (1998) Causal layered analysis: Poststructuralism as method. *Futures* 30 (8), 815–829.
Ivanov, S. and Stavrinoudis, T.A. (2018) Impacts of the refugee crisis on the hotel industry: Evidence from four Greek islands. *Tourism Management* 67, 214–223.
Khazaal, N. (2021) Introduction. In N. Khazaal and N. Almiròn (eds) *Like an Animal: Critical Animal Studies Approaches to Borders, Displacement, and Othering* (pp. 1–47). Brill.
Khazaal, N. and Almiròn, N. (eds) (2021) *Like an Animal: Critical Animal Studies Approaches to Borders, Displacement, and Othering*. Brill Rodopi.
Kline, C. (2018a) *Animals, Food, and Tourism*. Routledge.
Kline, C. (2018b) *Tourism Experiences and Animal Consumption: Contested Values Morality and Ethics*. Routledge.
Kline, C. (2021) *Socializing Animal-Based Tourism*. Routledge.
Kline, C. and Fischer, B. (2023) Morality on holiday: Inspiring ethical behaviour in animal-based tourism through non-moral values. *Tourism Recreation Research* 48 (2), 147–158.
Knight, J. (2009) Making wildlife viewable: Habituation and attraction. *Society & Animals* 17 (2), 167–184.
LeBoeuf, B.J. and Campagna, C. (2013) Wildlife viewing spectacle: Best practices from elephant seal (Mirounga sp.) colonies. *Aquatic Mammals* 39 (2), 132–146.
Marsi, F. (2023, Oct. 20) Italy's Lampedusa caught between solidarity and survival as migration rises. *Al Jazeera*. https://www.aljazeera.com/news/2023/10/20/italys-lampedusa-caught-between-solidarity-and-survival-as-migration-rises (accessed January 13, 2025).
Massey, D. (2005) *For Space*. Sage.
McIntosh, D. and Wright, P.A. (2017) Emotional processing as an important part of the wildlife viewing experience. *Journal of Outdoor Recreation and Tourism* 18, 1–9.
Merriman, P. (2012) *Mobility, Space, and Culture*. Routledge.
Mitchell, F. (1968) The economic value of wildlife viewing as a form of land use. *East African Agricultural and Forestry Journal* 33 (sup. 1), 98–103.
Mlakar, N. and Korže, S.Z. (2022) New tourism trend: Travelling with pets of pet sitting at a pet hotel? *Contemporary Issues in Tourism* 221, 221–242.
Pappas, N. and Papatheodorou, A. (2017) Tourism and the refugee crisis in Greece: Perception and decision-making of accommodation providers. *Tourism Management* 63, 31–41.
Rose, G. (2007) *Visual Methodologies: An Introduction to the Interpretation of Visual Materials* (2nd edn). Sage.
Smith, D.L. (2012) *Less than Human: Why We Demean, Enslave, and Exterminate Others*. Macmillan Publishers.
Smith, D.L. (2021) *Making Monsters: The Uncanny Power of Dehumanization*. Harvard University Press.
Taylor, M., Grimwood, B.S.R. and Boluk, K. (2018) Caring for animal welfare: Volunteer tourists and captive-elephant wellbeing. In B.S.R. Grimwood, H. Mair, K. Caton and M. Muldoon (eds) *Tourism and Wellness: Tourism for the Good of All?* (pp. 71–94). Lexington Books.

Tomassini, L. (2022) Can I bring my pet?: The space for companion animals in hospitality and tourism. *Research in Hospitality Management* 12 (2), 99–101.

Trilling, D. (2023, May 29) One EU policy the Tories are happy to emulate: Cracking down on refugees. *The Guardian*. https://www.theguardian.com/commentisfree/2023/may/29/eu-policy-tories-happy-to-emulate-crackdown-refugees

Waitt, G.R. (2005) Doing discourse analysis. In I. Hay (ed.) *Qualitative Research Methods in Human Geography*. Oxford University Press.

Wu, H.C. and Chang, Y.Y. (2023) Pet attachment, experiential satisfaction, and experiential loyalty in medical tourism for pets. *Tourism Recreation Research* 48 (2), 159–172.

Zenker, S., von Wallpach, S., Braun, E. and Vallaster, C. (2019) How the refugee crisis impacts the decision structure of tourists: A cross-country scenario study. *Tourism Management* 71, 197–212. https://10.1016/j.tourman.2018.10.015

8 Encountering the 'Other': The Role of Underwater Photographers in Shaping the Futures of Regenerative Marine Tourism

Hindertje Hoarau-Heemstra and Anne Wally Ryan

Introduction

Diving into the depths of the underwater world, this book chapter discusses the role of amateur underwater photographers in establishing a connection between humans and marine life, and thereby providing a foundation for the future regenerative tourism. Although regenerative tourism encourages both travellers and local communities to engage in sustainable practices that foster a sense of responsibility for the well-being of a destination (Ateljevic & Sheldon, 2022; Dredge, 2022), it is still unclear who is included in the vision of a positive feedback loop between tourism and the local community. Similarly, questions about how – and for whom – these regenerative practices are implemented often reflect a human-centred approach. In the last few years, many researchers have raised critical considerations regarding the focus and language used to describe the United Nations Sustainable Development Goals (SDGs), suggesting that their implementation reflects an underlying value set of 'humans above nature' (Boluk *et al.*, 2019; Kline *et al.*, 2022). Conversely, a tourism destination can be viewed as a multispecies community inhabited by species living on and deriving sustenance from the land, water, and air (Hoarau & Kline, 2021). The research context encompasses the coastal zone, which is commonly defined as the area that ranges from land directly influenced by sea or sea-related activities to one nautical mile outside the baseline (Ministry of Local Government and Regional Development, 2020).

Drawing from posthumanist perspectives (Wolfe, 2009; Haraway, 2016), this research challenges traditional views, emphasizing interconnectedness and relationality and redefining agency through interconnected entanglements (Alaimo, 2016). This approach prompts a re-evaluation of human-environment relationships and temporality (Cielemęcka & Daigle, 2019). Human impacts on marine ecosystems in the Anthropocene era have underscored the urgent need to understand and protect marine life (Hughes *et al.*, 2017). How can we understand the nonhuman otherness of marine life, connect with it, and protect it? Quoting Richard Louv, we acknowledge that 'we cannot protect something we do not love, we cannot love what we do not know, and we cannot know what we do not see. And touch. And hear' (2012: 104). Underwater photography has emerged as a powerful tool that inspires travel and fosters connections with natural environments; it has the potential to serve as a catalyst for positive change in marine tourism by inspiring appreciation, raising awareness, and prompting action to safeguard and rehabilitate unique underwater ecosystems for the benefit of present and future generations (Fennell & Yazdan Panah, 2020). However, whether marine wildlife tourism will exist in 2050 remains uncertain, as holograms, digital immersion, haptic technologies and gestural interfaces blur the boundaries between reality and science fiction (Yeoman & Postma, 2014). As a result, it is urgent to consider the future existence of thriving marine environments, where humans can respectfully engage with the nonhuman otherness of the marine life inhabiting the underwater space.

Hence, drawing on the notion of regenerative tourism – which posits the human capability to safeguard destinations and ensure their vitality and richness for future generations (Dredge, 2022) – this chapter addresses the following question: *how does underwater photography contribute to regenerative marine tourism futures?* To address this question, we take a posthuman stance to explore the role of amateur underwater photographers in fostering connections between humans and marine life – conceived here as a nonhuman otherness inhabiting the underwater space. By examining the interactions between amateur photographers and marine life, this study aims to inspire appreciation and action for sustainable development and regenerative tourism for all beings that affect or are affected by tourism. Utilizing the ontological futures classification system (Bergman *et al.*, 2010), we conduct a qualitative study of two protected marine tourism destinations and examine the current state of diving tourism and its possible futures. Our analysis focuses on the interactions between human and nonhuman actors in tourism, as well as the practices, emotions and images employed to represent the otherness of marine life.

Conceptual Framework

The era of the Anthropocene is characterized by the omnipresent influence of *Homo sapiens* and the impacts of human behaviour on a global scale (Hughes *et al.*, 2017). Despite the fact that human activities have left almost no corner of the world untouched (Bonneuil & Fressoz, 2016), certain wilderness regions, such as the Arctic and remote archipelagos, remain relatively inaccessible or challenging to inhabit. These areas, characterized by their pristine wilderness, attract tourists seeking uncommon experiences, and an increasing number of attractions are found beneath the surface of the water, indicating the growing popularity and exploration of marine tourism.

However, the development of tourism presents a dilemma: while tourism has the potential to conserve wilderness areas, it is also associated with increased accessibility and heightened human pressure, thereby diminishing the wild and remote nature of these destinations. Regenerative sustainability has emerged as a potential strategy for reconciling this paradoxical situation (Bellato *et al.*, 2023; Dredge, 2022). Consequently, regenerative tourism can be seen as a possible strategy for the future, offering a direction for development.

We acknowledge the emphasis of future studies on envisioning diverse future scenarios rather than prescribing a singular outcome (Inayatullah, 2002). Traditionally rooted in forecasting, futures studies have transitioned towards scenario planning, facilitating the exploration of multiple outcomes and guiding decision-making amidst uncertainty (Strand, 1999), which underscores the importance of embracing a plurality of futures and exploring various possibilities rather than seeking a definitive truth (Bergman *et al.*, 2010; Robertson & Yeoman, 2014; Yeoman & Postma, 2014). Forecasts provide insights into potential futures, offering insights into the paths we may choose to take as individuals and societies. Yeoman *et al.* (2012) distinguish 'signposts' as representing factual occurrences and evidence of scenarios occurring in the present. On the other hand, 'signals' serve as indicators of what the future could hold (Robertson & Yeoman, 2014). To analyze signposts and signals in the context of marine tourism, we outline potential drivers of future development drawn from the literature. By integrating insights from posthumanism and considering the role of tourism photography in regenerative sustainability, our framework aims to provide a holistic approach for understanding and shaping the future of remote marine tourism destinations.

Spatial and Temporal Multispecies Connections: A Posthuman Approach to Marine Tourism

Posthumanism is a philosophical and cultural movement (Alaimo, 2012; Haraway, 2016, 2003; Wolfe, 2009) that critiques and moves beyond traditional humanist concepts, including Cartesian dualism, which separates mind and body. It also challenges other binary oppositions, such as male/female, human/animal, and human/machine. Instead, posthumanism emphasizes the interconnectedness and fluidity of these categories, suggesting that the boundaries between them are not fixed, but rather mutable and contingent. Bennett *et al.* (2010) and Haraway (2016) advocate for the idea that all bodies are kin, forming a dense network of relations where harm to one part is harm to the whole. Wolfe (2009) understands differences between species as a force of becoming, where at the intersections of species, humans and others mutually transform and become. Hence, coexistence and interactions continuously shape all beings in the process of self-realization (Luhmann, 2012; Maturana & Varela, 1980). When defining our humanity, we artificially draw a boundary between our bodily systems and our environment. As this boundary is subjective, our definitions of what is outside, that is, otherness, reflect what we think about ourselves, our values and our ways of being in the world. Alaimo (2012) emphasizes viewing human existence as being intertwined with the world, prompting a reassessment of temporality and notions of futurity (Cielemęcka & Daigle, 2019). Hence, posthumanism as a framework has significant implications for understanding the future. Cielemęcka and Daigle (2019) proposed that sustainable development, from a posthumanist perspective, requires recognizing multispecies interconnectedness and moving away from a linear, human-centric vision of futurity. In times of environmental crises, humans feel compelled to consider the consequences of compromised interconnectedness. Interconnectedness spans temporal connections – past, present and future – shaping potential futures.

Emirbayer and Mische (1998) argue that agency can be seen as a temporally embedded process of social engagement shaped by the past and oriented towards imagining alternative possibilities for the future while addressing present concerns. Regarding concerns for the future, de la Bellacasa (2017) argues that care encompasses the actions necessary to sustain life and its diversity. Violence arises whenever 'our' future is deemed more significant or worthy of protection than that of other beings (Cielemęcka & Daigle, 2019). Therefore, envisioning the future as emerging from care for the entanglements of beings can foster an understanding of the practices needed to ensure regenerative tourism for both today and tomorrow.

Underwater Wildlife Photography for Regenerative Tourism Development

Based on the framework presented above, we assume that place development and preserving critical ecosystems require a holistic and critical approach to sustainable tourism that considers and addresses the multiple interconnected aspects and stakeholders involved in the tourism industry to achieve long-term environmental, sociocultural and economic sustainability. Hughes *et al.* (2017) define a marine protected tourist destination as an underwater space that attracts humans who respectfully engage with and photograph complex nonhuman otherness; this type of destination employs a marine life management approach based on transformed attitudes, knowledge, and governance frameworks. Hence, tourism should focus not only on attracting visitors but also on ensuring that the destination's natural and cultural assets are preserved, that there are benefits for local communities, and that tourists have meaningful and authentic experiences. The aim of regenerative tourism is to balance local community interests with tourism development (Dredge, 2022). However, we follow the argument of Boluk *et al.* (2019) that a 'critical consciousness' must be adopted in considering the idea of regenerative tourism. For whom is this approach regenerative? What about oppressed groups in society whose voices are not always represented in tourism development debates? Jamal and Stronza (2009) emphasize the importance of considering nature as a key stakeholder in sustainability that requires a place at the table. Regenerative tourism should, therefore, adopt a multispecies perspective and a posthumanism framework to include and benefit the more-than-human communities that inhabit tourism destinations (Hoarau-Heemstra & Kline, 2022; Kline *et al.*, 2023). Hughes *et al.* (2017) argued that humans must respectfully engage with nonhuman animals and that the complexity and management of marine life requires a transformation of attitudes, knowledge and governance. This transformation is needed for tourists, locals and government officials. Mei Pung *et al.* (2020) propose a conceptual model to understand tourist transformation whereby transformative tourism can result in a positive change in attitudes and values among those who participate in the tourism experience.

Hence, tourist transformation is facilitated through interactions with a destination's attributes and activities. This mainly occurs when there is a significant perception of difference, which leads to reflection on the old self and the emergence of a changed or new self. As memory and interpretation are key dimensions in transforming attitudes and knowledge, images can be a powerful instrument in the tourist transformation process. Photography can enhance engagement with the other by fostering an emotional connection between the observer and the observed. Interacting with photographs, as attributes of a destination, could be an opportunity for transformative learning (Mei Pung, 2020; Walter, 2016). For example, conservation photography documents environmental threats and seeks to increase awareness and proconservation behaviours (DeLuca & Demo, 2001). Emotional attachment to nature is an approach to increasing engagement with biodiversity (Hanisch *et al.*, 2019), as it increases viewers' feelings of kinship with animals and their perceptions of animals as individuals with personalities (Amiot & Bastian, 2015; Kalof *et al.*, 2011).

In addition to setting in motion transformative learning and emotional attachment, tourism photography can motivate the act of travelling as part of the tourism production system (Fennell & Yazdan Panah, 2020) and, as such, can open opportunities for playfulness, the alleviation of poverty, intercultural exchange, and togetherness (Markwell, 1997; Scarles, 2012). However, attracting large numbers of visitors to natural areas can have negative consequences when visitors are uninformed or misinformed or behave in harmful ways. The harmful behaviors of tourists who are wildlife photographers, such as approaching or baiting animals for the purpose of photographing them, can disrupt ecosystems (Green & Higginbottom, 2000; Lott, 1988). Ethical practices are therefore essential to ensure that wildlife photographers contribute positively to conservation and biodiversity protection. Timing is important for ethical wildlife photography, and codes of ethics, as voluntary mechanisms, must have a central place in educational regimes for ethical tourism; such considerations reflect the new wave of concern for animal ethics that permeates tourism scholarship and practices around respect, dignity, and the reduction of suffering (Bertella, 2013; Fennell & Yazdan Panah, 2020; Kline *et al.*, 2023). 'Do no harm' should be the primary objective of the wildlife photographer, along with a deliberate attempt to induce others to protect habitats, raise conservation funds and increase awareness (Mittermeier *et al.*, 2005). However, clear codes of ethics must be established given the unique challenges of wildlife photography. This is even more valid for wildlife photography in marine environments, where practices are literally hidden below the surface.

We see both the development of a new code of ethics and changes in attitudes and values as necessary steps towards the realization of regenerative marine tourism destinations. They are signposts on the road towards a more regenerative, fair and sustainable future tourism, which we would argue is a preferred future for tourism (Robertson & Yeoman, 2014).

Methodology

In our pursuit of evidence for potential futures, we explore signals and signposts within the context of our scenario. To understand what potential futures might look like and how they are categorized, we need a framework that examines the essence and existence of these futures based on the present. To understand the present, we employed a narrative methodology involving unstructured interviews within an open dialogic setting (Tomassini *et al.*, 2024).

In this study, we engaged with dive operators, dive instructors and underwater photographers and participated in diving and snorkelling practices. Our fieldwork involved observing various tourism experiences, adhering to ethical guidelines outlined by The National Committee for Research Ethics in the Social Sciences and the Humanities (NESH, 2022). The interviews were conducted with informed consent, ensuring anonymity of all participants. The data collected included interview notes and transcripts, fieldwork journals, photographs, social media posts and published documents. Our analysis followed an open coding approach, identifying practices related to interactions between humans and other-than-human beings in marine protected areas. Specifically, we focused on amateur underwater photographers as the human subjects and fish inhabiting marine protected areas frequently visited by divers and photographers as the nonhuman subjects.

Marine protected area	Fieldwork period	Interviews (formal and informal)	Observations
Saltstraumen, Norway	May 2022	4	3
Gardens of the Queen, Cuba	November 2022	5	7

While mammals share similarities with humans in terms of living on land and raising offspring, connecting with fish, as non-land-based creatures, is more challenging, but it is essential for species and habitat conservation. We delved into the narratives of the first author, who conducted the ethnography, capturing their first-hand encounters with human-marine animal interactions in two marine protected areas: Saltstraumen, Norway and Gardens of the Queen, Cuba. This ethnographic approach resulted in a scenario of regenerative tourism development in the context of marine protected areas. Examples were grouped into categories representing practices to allow for the identification of the general patterns and viewpoints expressed by our informants. The following themes emerged: photography and consideration of animal ethics, the caring underwater observer and achieving a balance between nature and society. We present our research context below.

Context of the research

The geographical context of this study is the coastal zone, more specifically focusing on marine protected areas as the loci of interaction between species and the role of underwater photography in caring for the other species. We have chosen two locations that vary in terms of type of tourism, climate, governance, culture, and nature: Saltstraumen in Norway and Gardens of the Queen in Cuba.

Saltstraumen, a small strait with one of the strongest tidal maelstrom currents in the world, is located in the municipality of Bodø in Nordland County, Northern Norway. Powerful tidal currents play a crucial role in local marine ecosystems, which have rich biodiversity. The area is home to the wolffish, a notable species in those northern waters. The wolffish, known for its robust build and distinctive appearance, is a cold-water fish found in the North Atlantic and Arctic Oceans. The Saltstraumen's currents provide a suitable habitat for these fish, which find shelter in the rocky crevices and underwater formations. Their preference for colder waters makes this strait an ideal location for their survival. These fish, which are characterized by elongated bodies and strong jaws filled with formidable teeth, play an integral role in the local marine ecosystem (Falk-Petersen *et al.*, 2010). As predators, they help maintain ecological balance by controlling populations of smaller fish and invertebrates, contributing to the overall health of the underwater environment (Bluemel *et al.*, 2022). Despite being less known than other marine creatures, the wolffish is a crucial component of the biodiversity within the Saltstraumen. Their presence contributes to the allure of the area for divers, wildlife photographers, and tourists seeking glimpses of the unique marine life thriving in its powerful currents.

Saltstraumen has attracted both conservation efforts and tourism for several years; however, it is also a primary resource for settlements. The remains of a 10,000-year-old hunter settlement in the area are the oldest known traces of human settlement in Bodø; the settlement was also one of the oldest archaeological discoveries in Norway (Wikipedia, 2023). Saltstraumen's unique natural features attract tourists, for example, it offers activities such as diving, wildlife photography, fishing and rib-safaris (Borch, 2004). Saltstraumen has been named a marine protected area (MPA, anchored in the Norwegian legal system), and it was selected as one of The Wild Wonders of Europe in a collaboration between the National Geographic and the WWF.

Gardens of the Queen, known as Jardines de la Reina in Spanish, is a pristine archipelago located off the southern coast of Cuba in the Caribbean Sea. Cuba, the largest archipelago in the Caribbean, boasts of some of the most well-conserved marine environments in the Western Atlantic. It encompasses vast stretches of pristine coastal and ocean habitats, which, in certain regions, experience minimal human-induced pressures and harbor a rich variety of fish species (Goulart *et al.*, 2018; Angulo-Valdes *et al.*, 2022). Gardens of the Queen is renowned for its coral reefs, mangrove forests, seagrass beds, and diverse marine life. The protected marine park spans more than 2200 square kilometres and is considered one of the healthiest and most vibrant marine ecosystems in the Caribbean. The marine reserve network of Cuba covers an area equivalent to 61% of the total protected area within the Caribbean biodiversity hotspot (Goulart *et al.*, 2018). The area is rich in marine biodiversity, making it a haven for various species. Among these is the goliath grouper, a massive and critically endangered fish that inhabits the Gardens of the Queen. The goliath grouper is one of the largest species of grouper in the Atlantic Ocean, reaching lengths of up to 8 feet (2.4 m) and weighing several hundred pounds (Koenig *et al.*, 2020). The fish has an ideal habitat within the protected waters of the Gardens of the Queen. The area's strict conservation efforts and limited human impact have allowed these majestic creatures to thrive. Goliath groupers play a crucial role in the ecosystem as apex predators, maintaining the balance of the underwater food chain.

Goliath groupers offer a unique diving experience that has captured the attention of millions of divers from around the world (Koenig *et al.*, 2020). Visitors to Gardens of the Queen can engage with the underwater ecosystem through diving and snorkelling tours; however, these are strictly regulated by the Cuban authorities. Diving operations and conservation efforts have been challenged since the pandemic and have been further complicated by the socialpolitical situation in Cuba. Research on tourism in Cuba highlights sociocultural impacts, revealing tensions between the revolutionary ideals of socialism and the growing tourism industry. While some Cubans acknowledge the economic benefits of tourism, such as higher incomes and improved living standards for those with direct access to tourists, there are growing concerns about food insecurity. Record tourism arrivals have reduced the availability of food for local consumption, leading to rising prices and the diversion of agricultural goods to the black market (Rutty & Richardson, 2019). The development of a sustainable tourism strategy for Cuba necessitates long-term considerations, equity, and collaboration among stakeholders to optimize economic, social, and environmental linkages. Therefore, there is a need for sustainable tourism planning and development in that region, with a focus on the local level (Knollenberg *et al.*, 2020). Rutty and Richardson argue that addressing Cuba's unique challenges, including bureaucratic hurdles and state market regulation, requires a comprehensive understanding of the environmental and sociocultural impacts of tourism. However, obtaining valid and reliable data can be challenging due to unclear authorization procedures and limited access to key tourism stakeholders. For instance, establishing contact with stakeholders outside Cuba can be difficult, as emails may be blocked. Moreover, foreign researchers may struggle to elicit honest viewpoints from tourism stakeholders, with concerns about intentional unreliability. Consequently, the research on Cubans' perceptions of tourism has primarily relied on participant observation or ethnography (Rutty & Richardson, 2019).

Diving Tourism in Marine Protected Areas

Human activities significantly degrade coastal environments worldwide, exacerbated by their hosting of nearly half of the world's population and rich biodiversity (Arcos-Aguilar, 2021). This degradation leads to a scarcity of marine biodiversity, making areas with abundant marine life attractive for nonextractive activities such as diving tourism. Diving tourism has grown from a niche activity to a flourishing industry, enticing enthusiasts with the opportunity to experience pristine wildlife encounters (Dai *et al.*, 2024). Marine wildlife photography requires not only photographic skills but also diving equipment, boats and guides, with scuba diving being a globally practiced high-risk sport since the 1970s. Dive clubs commonly require certification to ensure safety and predictability among divers. Once underwater, divers are guided by dive masters who communicate about safety and identify species using globally recognized sign language (Koenig *et al.*, 2020). Preserving the biodiversity of diving sites requires consideration of the long-term impacts on the environment, economy, and livelihoods of local communities. Given their economic significance, diving operators should actively participate in marine spatial planning to ensure community-based management (Arcos-Aguilar, 2021).

Findings: Utopian Scenarios of Regenerative Diving Tourism Destinations

The forward-looking perspective embraces the intricate dynamics and uncertainties of today's world. Rather than projecting the past and present into the future, this approach takes a different direction and projects the future into the present (Postma *et al.*, 2017). This involves identifying the driving forces of change and key uncertainties and then crafting scenarios that vividly depict plausible futures. Hence, using a scenario-planning methodology is a process of providing different scripts of the future in which the central focus is not certainty and prediction but the ability to support thinking and decision-making by imagining the possible and the impossible, thereby allowing participants to cocreate, own and make sense of the future. Based on the data analysis of our case study of Gardens of the Queen and Saltstraumen, the following future utopian scenarios were envisioned: *Regenerative diving destinations.*

Regenerative Diving Destinations: Gardens of the Queen, Cuba, and Saltstraumen, Norway, in 2050

Due to a grassroots movement led by diving tourism businesses, the marine protected areas of Gardens of the Queen and Saltstraumen have been preserved for future generations. These two seemingly distinct regions faced a triple crisis of climate change, pollution and biodiversity loss within their marine protected areas. Surprisingly, the connections created by divers and diving tourism businesses ignited a global wave of environmental activism.

Gardens of the Queen stands as a flagship example of successful conservation and marine management in the Caribbean. Once a private diving spot for Fidel Castro, this area nearly lost its wealth of marine life in the 2020s.

When international travel resumed after the pandemic, Cuba saw a surge in visitors. Cuba's location along the Caribbean cruise route exposed it to growing marine litter that threatened its reefs and their inhabitants. Moreover, on the other side of the globe, above the Arctic Circle, Saltstraumen faced similar challenges. Pollution from recreational fishing, sewage pollution in the fjord, and over-tourism resulting from easily accessible technology nearly eradicated all marine life. This area, once teeming with cod, halibut, wolfish, coalfish, and small species such as nudibranchs, was designated a marine protected area in 2009. However, fishing continued, targeting the largest fish. The marine debris from thousands of fishing lines further devastated the ecosystem, harming to the fish and birds ensnared in them. Uncontrolled sewage outlets near the marine protected areas introduced an excess of nutrients into once-clear waters.

The determination of local communities to preserve these areas led to a global campaign in 2035 demanding stricter and more nature-friendly marine management. Striking images shared worldwide became powerful tools for promoting marine tourism and educating the public about the beauty and importance of underwater ecosystems. Underwater photography, capturing the breathtaking beauty of marine environments and their inhabitants, increased awareness of ecosystem fragility. These compelling underwater images, focusing on portraits of key species such as the wolffish and goliath grouper, also played a crucial role in advocacy efforts to influence policymakers and drive changes in regulations and policies related to marine conservation and sustainable tourism. This heightened awareness translated into increased support for conservation initiatives.

Through this campaign, the communities in Saltstraumen and Cuba became connected and began collaborating. They exchanged ideas, advocating for degrowth and regenerative tourism practices in harmony with nature. By 2050, these areas have not only survived but also have rebounded to a state of health and abundance not seen in the past century. These examples inspired other communities to utilize their marine protected areas to enforce local sustainability and degrowth, aligning with ecological systems.

Underwater photography played a pivotal role in developing regenerative tourism by highlighting sustainable and regenerative practices. It showcased destinations and operators committed to minimizing their environmental impact and promoting responsible tourism. Involving local communities in underwater photography initiatives fostered a sense of ownership and responsibility for marine conservation. Additionally, it provided economic opportunities to local residents through activities such as guiding and photography services. A global code of ethics for underwater wildlife tourism has been developed by tourism operators in Saltstraumen and Cuba to represent different ecosystems and knowledge bases.

Despite the attention and availability of high-quality underwater images attracting tourists and divers to these destinations, stringent regulations, quotas, and higher prices boosted local economies. This economic incentive encouraged communities to prioritize the protection of their marine environments. Visual storytelling through underwater photography had a significant impact on people's attitudes and behaviors towards the marine environment, promoting responsible tourism practices, such as reef-friendly sunscreen use and responsible diving techniques.

More biodiverse marine areas proved to be resilient to climate change, offering numerous ecosystem services to communities. Underwater photographers often collaborated closely with marine biologists and scientists to document species, behaviour and ecological changes, providing valuable data for research and aiding in the understanding and preservation of marine life.

Discussion

Signals and signposts for change

The scenario we present depicts a series of significant changes compared to current diving tourism in marine protected areas. These changes arise from increasing environmental consciousness, the development of ethical guidelines for underwater photographers, connections and collaboration between regions, the worsening triple crisis of climate, biodiversity and pollution, the growing emphasis on species equality with marine life being granted rights, and the utilization of technology for interacting with marine species.

In our study, the signposts that are relevant to the transformation in how humans care for, manage, and interact within marine protected areas include the following: the spread of concerns regarding environmental ethics (Hoarau-Heemstra & Hjalager, 2020); the advancement of animal rights and the adoption of more ethical practices in marine protected area management (Fennell & Sheppard, 2021; Kline *et al.*, 2023); fishing and tourism; the continuous development of underwater photography technology and diving equipment; the integration of immersive technology; collaborative efforts between the realms of science, local communities, and governance; and the strengthening of international collaboration within marine national park networks. These signposts indicate that as marine protected areas evolve, they will witness emergent collaborations across sectors, the adoption of new technologies, and increased expectations from customers, driven by climate change and nature protection advocacy. The pertinent signals in this scenario herald the need for establishing and nurturing connections among stakeholders about the utilization and management of marine protected areas. Furthermore, these signals highlight the gradual shift in the patterns of nature-based experience consumption (Hjalager & Heemstra, 2020). New questions arise, particularly with regard to how posthumanist and new-materialist perspectives and values propagate from pioneering individuals and organizations to wider tourism stakeholder networks in pursuit of reaping further benefits (Tomassini *et al.*, 2024).

Diving and taking pictures: From hedonistic experience to responsible practice

Why do we take pictures as part of the tourism experience? The reasons for engaging in underwater experiences, either in the context of tourism or leisure, are often driven by the search for pleasure, sensations, and extraordinary experiences. Underwater images are used for exhibiting on social media, in photography contests, or in physical exhibitions. Recently, Zhan *et al.* (2024) highlighted the impact of external influences, such as media coverage and images aimed at addressing animal abuse within the tourism industry, on tourists' cognitive judgement regarding animal ethics. We also argue that media coverage using underwater images is an influential factor in the context of animal-related tourism and that emotional responses to animal ethics and nature protection will be transformed. Hence, underwater images change the future because they change the way people feel and think about marine animals and environments.

However, from our scenario, it becomes clear that underwater photography is not without risk to the marine protected area and its inhabitants (Fennell & Yazdan Panah, 2020). Taking pictures can impact marine life because the photographer may touch a reef to maintain their balance while taking a picture, flashlights might scare or disturb underwater animals, and the desire to capture a good image of an animal might result in harassment of the animal. To develop regenerative tourism (Dredge, 2022) in marine protected areas, wildlife photography must be subjected to ethical guidelines and regulations (Fennell & Yazdan Panah, 2020). What is allowed, by whom, when, and how are questions that must be answered for each diving or marine tourism destination? Depending on the sensitivity of the individual species, we propose that photographers follow local rules and regulations for ethical photography practices. To meet these guidelines, the people who work most with or depend most on the marine protected area, such as the local community, indigenous peoples, tourism operators and policymakers, should aim to collaborate and cocreate such destination-specific guidelines. This code of conduct must be communicated and controlled by local stakeholders to ensure that photographers are both aware of and respect the wishes of locals on how these practices are carried out in their local multispecies communities.

The caring observer underwater

Interaction and connection with other species of animals through eye contact and behavior can build bonds and evoke care practices (Alaimo, 2016). In tourism, diving equipment, underwater photography gear, and knowledge of the underwater climate, landscape and currents are provided by diving operators and guides who know the area and its inhabitants well. These diving masters and guides can be seen as gatekeepers of care and ethical behaviors (Dai *et al.*, 2024). There are few places where divers can go without a guide, and even fewer people carry their own compressors and tanks with them while travelling. This means that almost all divers are in contact with a diving club before going underwater. Our data show that in Saltstraumen, divers are not allowed to take pictures of wolffish that are guarding their eggs in a cave, touch wolffish or hunt wolffish. In the Gardens of the Queen, the guides tell the story of the goliath grouper and how it is critically endangered. However, these fish are very curious and come close to divers, resulting in people trying to touch them. These large groupers are also able to communicate to divers and guides where lobsters are hiding in the rocks, and thus, humans help to lure the lobsters out. This kind of multispecies collaboration in hunting one species for the sake of another goes further than just observing marine wildlife (Kline *et al.*, 2023). The practice of feeding is popular among photographers because it results in actions and behaviors that produce effective and compelling photos and videos. However, local diving communities must ask critical questions and decide whether these practices help or harm the ecosystem, and this should be part of the ethical guidelines and interpretation that every guide follows (Fennell & Yazdan Panah, 2020).

Balance between nature and society: How to achieve it?

Philosopher Mark Rowlands' (2009) book The Philosopher and the Wolf reveals how sharing our living spaces with nonhuman others can impact our understanding of ourselves and the world. It is a recognition of the occurrence – even the necessity of trans-species communication. The animals we live with or encounter are not passive receptacles but themselves have agency; they respond to us, and we respond to them in multiple ways (Rowlands, 2009). Do humans stand outside the marine ecosystem, and where are the borders of these ecosystems? Additionally, the conservation of marine protected areas is a social and political act, and education is a social and political act. From whom are these multispecies areas protected, and can the principles of conservation and care through use be further developed for a future perspective that balances the needs of nonhumans and humans?

The concept of regenerative tourism (Bellato *et al.*, 2023) encourages underwater photographers to be informed and engaged in sustainable practices, fostering a sense of responsibility and the desire to make a difference. In other words, active agencies can better develop tourism in marine protected areas. This may indicate a shift from producing images for the purpose of sharing on social media or participating in photography contests or physical exhibitions mainly for one's own experience and recognition, to a consciousness of using these platforms to advance respectfulness, improve attitudes and facilitate learning about the encountering of others. In doing so, underwater photographers also gain responsibility, respect, and empowerment and can be included and collaborate in the management of marine protected areas.

The interests of park agency managers and conservation scientists may be oriented predominantly towards ecological integrity and conservation from 'Western' scientific perspectives, but local people living within protected areas also have valuable knowledge to contribute, that is, local/traditional knowledge (Jamal & Stronza, 2009). To influence people, it helps to understand people. In other words, residents, managers and tourist providers are not a homogenous group of 'locals'. Moreover, underwater photographers have their own unique characteristics; they are humans with individual personalities and ways of being. The next step in the collaboration among this diverse group of people involves caring for MPAs and coastal areas and balancing sustainable living areas for both nonhumans and humans. A growing body of research is emerging that rethinks the human/nonhuman binary; this area of research seeks to combine a scientific (biophysical) understanding of nature and environmental systems with a social science approach addressing complex issues faced by society (Tomassini *et al.*, 2024; Hoarau-Heemstra & Kline, 2022; Egner & Elverfeldt, 2009), such as food production, tourists' access to marine areas, and climate change. As stakeholders, amateur underwater photographers can promote a different understanding of the planet and our place within it, a different way of relating to and valuing nature.

Conclusion: The Futures of Regenerative Marine Tourism

In our future scenario and analysis, we consider several signals and signposts that can either develop tourism destinations in a neoliberal capitalist direction where marine life is seen as a resource for economic gain for businesses or toward a regenerative state where local ecosystems are cared for by all actors, both human and other-than-humans, for the sake of the local multispecies community. The relationship between locals and visitors must be considered in developing regenerative approaches to tourism in marine areas. Locals' views regarding tourism in their community and what they expect from their guests can be communicated in a range of ways such as through visitor guidelines and guided discussions about life above and under water that include specific examples. Communication and exchanges between regions that face similar challenges are essential to facilitate learning and innovation (Hoarau & Kline, 2014). It ensures the spread of good practices and knowledge across the globe. Furthermore, knowledge of the natural sciences and arts/humanities and social experiences of nature should be included as sources of knowledge in the management of marine protected areas. What these areas mean to humans and other beings should guide rules and regulations.

Our discussion shows the importance of what photographers decide to show and to whom they reach out. Photography is embedded in social and political power and guides the direction of public opinion. As highlighted in the current study, local communities can tell their stories to the world via social media and physical photo exhibitions. The sharing of images and alternate perspectives is also a way of giving back to the local community, in this case by facilitating their learning about and attaining a deeper understanding of the underwater part of their home. To be regenerative, the governance of marine protected areas should be transparent and inclusive. Self-governance in the form of ethical guidelines should be an integral part of management to ensure that practices in marine protected areas are based on connections between species and carried out in a caring and responsible way.

References

Arcos-Aguilar, R., Favoretto, F., Kumagai, J.A., Jiménez-Esquivel, V., Martínez-Cruz, A. L. and Aburto-Oropeza, O. (2021) Diving tourism in Mexico–Economic and conservation importance. *Marine Policy* 126, 104410.

Alaimo, S. (2012) Sustainable this, sustainable that: New materialisms, posthumanism, and unknown futures. *Pmla* 127 (3), 558–564.

Alaimo, S. (2016) *Exposed: Environmental Politics and Pleasures in Posthuman Times*. University of Minnesota Press.

Angulo-Valdes, J., Pina-Amargos, F., Figueredo-Martin, T., Fujita, R., Haukebo, S., Miller, V. ... and Whittle, D. (2022) Managing marine recreational fisheries in Cuba for sustainability and economic development with emphasis on the tourism sector. *Marine Policy* 145, 105254

Ateljevic, I. and Sheldon, P.J. (2022) Guest editorial: Transformation and the regenerative future of tourism. *Journal of Tourism Futures* 8 (3), 266–268.

Amiot, C.E. and Bastian, B. (2015) Toward a psychology of human–animal relations. *Psychological Bulletin* 141 (1), 6.

Bellato, L., Frantzeskaki, N. and Nygaard, C.A. (2023) Regenerative tourism: A conceptual framework leveraging theory and practice. *Tourism Geographies* 25 (4), 1026–1046.

Bennett, N., Lemelin, R.H., Johnston, M. and Nation, L.K.E.D.F. (2010) Using the social economy in tourism: A study of National Park creation and community development in the Northwest Territories, Canada. *Journal of Rural and Community Development* 5 (1).

Bergman, A., Karlsson, J.C. and Axelsson, J. (2010) Truth claims and explanatory claims. An ontological typology of futures studies. *Futures* 42 (8), 857–865.

Bluemel, J.K., Fischer, S.H., Kulka, D.W., Lynam, C.P. and Ellis, J.R. (2022) Decline in Atlantic wolffish Anarhichas lupus in the North Sea: Impacts of fishing pressure and climate change. *Journal of Fish Biology* 100 (1), 253–267.

Boluk, K.A., Cavaliere, C.T. and Higgins-Desbiolles, F. (2019) A critical framework for interrogating the United Nations Sustainable Development Goals 2030 Agenda in tourism. *Journal of Sustainable Tourism* 27 (7), 847–864. https://doi.org/10.1080/09669582.2019.1619748

Borch, T. (2004) Sustainable management of marine fishing tourism. Some lessons from Norway. *Tourism in Marine Environments* 1 (1), 49–57.

Bonneuil, C. and Fressoz, J.B. (2016) *The Shock of the Anthropocene: The Earth, History and Us*. Verso Books.

Cielemęcka, O. and Daigle, C. (2019) Posthuman sustainability: An ethos for our anthropocenic future. *Theory, Culture & Society* 36 (7–8), 67–87.

Dai, Y.D., Huang, F.H., Chen, K.Y., Chen, W.J. and Huan, T.C. (2024) Recreational scuba-diving as a special form of tourism: Lessons from Taiwan. *Tourism Review* 79 (3), 757–779.

de La Bellacasa, M.P. (2017) *Matters of Care: Speculative Ethics in More Than Human Worlds* (Vol. 41). U of Minnesota Press.

DeLuca, K.M. and Teresa Demo, A. (2001) Imagining nature and erasing class and race: Carleton Watkins, John Muir, and the construction of wilderness. *Environmental History* 6 (4), 541–560.

Dredge, D. (2022) Regenerative tourism: Transforming mindsets, systems and practices. *Journal of Tourism Futures* 8 (3), 269–281.

Egner, H. and Von Elverfeldt, K. (2009) A bridge over troubled waters? Systems theory and dialogue in geography. *Area* 41 (3), 319–328.

Emirbayer, M. and Mische, A. (1998) *What is agency? American Journal of Sociology* 103 (4), 962–1023. https://doi.org/10.1086/231294

Falk-Petersen, I.B., Kanapathippilai, P., Primicerio, R. and Hansen, T.K. (2010) Size, locality and seasonally related feeding preferences of common wolffish (Anarhichas lupus L.) from north-Norwegian waters. *Marine Biology Research* 6 (2), 201–212.

Fennell, D.A. and Yazdan Panah, H. (2020) Tourism and wildlife photography codes of ethics: Developing a clearer picture. *Annals of Tourism Research* 85, 103023.

Fennell, D.A. and Sheppard, V. (2021) Tourism, animals and the scales of justice. *Journal of Sustainable Tourism* 29 (2–3), 314–335.

Fennell, D.A., Kline, C., Mkono, M., Grimwood, B.S., Sheppard, V.A., Dashper, K. ... and Madrid, R. (2024) Tourism, animals & the vacant niche: A scoping review and pedagogical agenda. *Current Issues in Tourism* 27 (22), 3820–3848. https://doi.org/10.1080/13683500.2023.2280704

Goulart, F., Galán, Á.L., Nelson, E. and Soares-Filho, B. (2018) Conservation lessons from Cuba: Connecting science and policy. *Biological Conservation* 217, 280–288.

Green, R.J. and Higginbottom, K. (2000) The effects of non-consumptive wildlife tourism on free-ranging wildlife: A review. *Pacific Conservation Biology* 6 (3), 183–197.

Green, R. and Higginbottom, K. (2001) Negative effects of wildlife tourism and wildlife. *Wildlife Tourism Research Report Series n. 5*. Cooperative Research Centre for Sustainable Tourism.

Hanisch, E., Johnston, R. and Longnecker, N. (2019) Cameras for conservation: Wildlife photography and emotional engagement with biodiversity and nature. *Human Dimensions of Wildlife* 24 (3), 267–284.

Haraway, D.J. (2003) *The Companion Species Manifesto: Dogs, People, and Significant Otherness* (Vol. 1, pp. 3–17). Prickly Paradigm Press.

Haraway, D.J. (2016) *Staying with the Trouble: Making Kin in the Chthulucene*. Duke University Press.

Higgins-Desbiolles, F. (2025) On what terms might regenerative tourism proponents justly engage with indigenous knowledges? *Journal of Travel Research* https://doi.org/10.1177/00472875251341308

Hoarau, H. (2014) Knowledge acquisition and assimilation in tourism-innovation processes. *Scandinavian Journal of Hospitality and Tourism* 14 (2), 135–151.

Hoarau-Heemstra, H. and Hjalager, A.M. (2020) Safeguarding sustainable futures for marine wildlife tourism through collaboration and innovation: The utopia of whale-watching. In G. Bertella (ed.) *Wildlife Tourism Futures: Encounters with Wild, Captive and Artificial Animals* (pp. 113–125). Channel View Publications.

Hoarau, H. and Kline, C. (2014) Science and industry: Sharing knowledge for innovation. *Annals of Tourism Research* 46, 44–61.

Hoarau-Heemstra, H. and Kline, C. (2022) Making kin and making sense of human-animal relations in tourism. *Ecological Economics* 196, 107396.

Hughes, T., Barnes, M., Bellwood, D. et al. (2017) Coral reefs in the Anthropocene. *Nature* 546, 82–90. https://doi.org/10.1038/nature22901

Inayatullah, S. (2002) Reductionism or layered complexity? The futures of futures studies. *Futures* 34 (3–4), 295–302.

Jamal, T. and Stronza, A. (2009) Collaboration theory and tourism practice in protected areas: Stakeholders, structuring and sustainability. *Journal of Sustainable Tourism*, 17 (2), 169–189.

Koenig, C.C., Coleman, F.C. and Malinowski, C.R. (2020) Atlantic Goliath Grouper of Florida: To fish or not to fish. *Fisheries* 45 (1), 20–32.

Knollenberg, W., Kline, C., Jordan, E. and Boley, B.B. (2020) Will US travelers be good guests to Cuba? Examining US traveler segments' sustainable behavior and interest in visiting Cuba. *Journal of Destination Marketing & Management* 18, 100505.

Kalof, L. and Montgomery, G.M. (eds) (2011) *Making Animal Meaning*. MSU Press.

Kline, C.S., Hoarau-Heemstra, H. and Cavaliere, C.T. (2023) Wildlife equity theory for multispecies tourism justice. *Journal of Travel Research* 62 (6), 1167–1180.

Lott, D.F. (1988) Feeding wild animals: The urge, the risks, and the consequences. *Anthrozoös* 1 (4), 255–257.

Louv, R. (2012) *The Nature Principle: Reconnecting with Life in a Virtual Age*. Algonquin Books.

Luhmann, N. (2012) *Theory of Society, Volume 1*. Trans. Rhodes Barrett. Stanford University Press.

Markwell, K.W. (1997) Dimensions of photography in a nature-based tour. *Annals of Tourism Research* 24 (1), 131–155.

Maturana, H.R. and Varela, F.J. (1980) *Autopoiesis and Cognition: the Realization of the Living*. Springer Nature Link.

Ministry of Local Government and Regional Development (2020) Planning in the sea area (only in Norwegian). Retrieved from regjeringen.no/no/dokumenter/planlegging-i-sjoomradene/id2700314/?ch=1

Mittermeier, R.A., Da Fonseca, G.A., Rylands, A.B. and Brandon, K. (2005) A brief history of biodiversity conservation in Brazil. *Conservation Biology* 19 (3), 601–607.

NESH (2022, 26.05) The National Committee for Research Ethics in the Social Sciences and the Humanities (NESH). Retrieved from forskningsetikk.no/en/about-us/our-committees-and-commission/nesh/guidelines-nesh/guidelines-for-research-ethics-in-the-social-sciences-humanities-law-and-theology2/

Postma, A., Cavagnaro, E. and Spruyt, E. (2017) Sustainable tourism 2040. *Journal of Tourism Futures* 3 (1), 13–22.

Pung, J.M., Gnoth, J. and Del Chiappa, G. (2020) Tourist transformation: Towards a conceptual model. *Annals of Tourism Research* 81, 102885. https://doi.org/10.1016/j.annals.2020.102885

Robertson, M. and Yeoman, I. (2014) Signals and signposts of the future: Literary festival consumption in 2050. *Tourism Recreation Research* 39 (3), 321–342.

Rowlands, M. (2009) *The Philosopher and the Wolf: Lessons from the Wild on Love, Death and Happiness*. Granta Books.

Rutty, M. and Richardson, R.B. (2019) Tourism research in Cuba: Gaps in knowledge and challenges for sustainable tourism. *Sustainability* 11 (12), 3340.

Scarles, C. (2012) The photographed other: Interplays of agency in tourist photography in Cusco, Peru. *Annals of Tourism Research* 39 (2), 928–950.

Strand, S. (1999) Forecasting the future: Pitfalls in controlling for uncertainty. *Futures* 31 (3–4), 333–350.

Tomassini, L., Bertella, G., Grasso, C. and Lenzi, C. (2024) The space of animal justice in wildlife sanctuaries: A posthuman perspective. *Journal of Ecotourism* 23 (1), 38–55.

Yeoman, I. and Postma, A. (2014) Developing an ontological framework for tourism futures. *Tourism Recreation Research* 39 (3), 299–304.

Yeoman, I., Robertson, M. and Smith, K. (2014) A futurist's view on the future of events. In S.J. Page and J. Connell (eds) *The Routledge Handbook of Events* (pp. 518–536). Routledge.

Walter, P.G. (2016) Catalysts for transformative learning in community-based ecotourism. *Current Issues in Tourism* 19 (13), 1356–1371.

Wikipedia (2023, October 31st) Saltstraumen. Retrieved from https://en.wikipedia.org/wiki/Saltstraumen.

Wolfe, C. (2009) Humanist and posthumanist antispeciesism. In P. Cavalieri (ed.) *The Death of the Animal: A Dialogue* (pp. 45–58). Columbia University Press. https://doi.org/10.7312/cava14544-004

Zhan, L., Cheng, M. and Zhu, J. (2024) Progress on image analytics: Implications for tourism and hospitality research. *Tourism Management* 100, 104798.

9 Video Methods in Co-Creating Space for Non-Human Animals in Tourism Futures

Mikko Äijälä, Minni Haanpää and
José-Carlos García-Rosell

Mobile Video Ethnography in the Field: Putting Booties on a Dog

It is late autumn and the first sub-zero temperatures have made the tracks rock hard. Hence, some of the dogs have to wear booties to protect their paws. I really struggle to get the booties on as the dog does not like my intentions at all. The dog is crouching, keeping its tail down and subtly trying to escape my attempts to put booties on its paws. I feel embarrassed because the mushers look at us. Moreover, the dog senses that I am nervous, which makes it even harder. When going back over the scene on the video with the guides I feel even more embarrassed. The guides educate me that my technique is wrong, as I should not try to lift my other leg over the dog, as you might imagine it is afraid of that (Figure 9.1).

Videography at a National Travel Fair: Disseminating Human-Animal Spaces

We are organising a panel discussion on ethical tourism business in relation to animals at the national Matka Travel Fair in Helsinki, Finland. As a part of the panel we are screening the 'Ethical Tourists' videography. I'm nervous about facilitating a panel at such a large industry event and particularly about showing the videography to the audience. The film starts and I see the audience concentrating on listening to our interviewees, but also noticing the scenes from the animal-based tourism activities. After the film a lively conversation begins on how the tourists' attitudes towards animals will shape tourism practices and the industry in the future (Figure 9.2).

Figure 9.1 A snapshot from a video of putting booties on a dog

Figure 9.2 Panel discussion at a national travel fair in which videography was used as a discussion stimulator

These two vignettes present the possibility that video-based methods hold to disrupt habituated ways of thinking and acting within animal-based tourism. The video has the ability to move, touch and affect us – from tourism practitioners at a local level to international audiences – beyond the boundaries of a particular space and the actual event at hand.

Introduction

Over the past decade, there has been a growing scholarly interest in examining the messy entanglements and relationships that are enacted during animal-based tourism encounters (Fennell, 2012; Rickly & Kline, 2021). This comes as no surprise, given the increasing popularity of animal-based tourism and the 'animal turn' that is affecting tourism among many other social science disciplines (Carr & Broom, 2018; Danby *et al.*, 2019). Within this animal-based tourism academia, some studies have acknowledged the presence and agency of non-human animals in shaping the practices, experiences and landscapes of tourism (e.g. Äijälä *et al.*, 2021; Bertella, 2014; Danby *et al.*, 2019). Despite these advancements and increasing awareness of animals within the tourism industry, our understanding of the role non-human animals play in these encounters remains somewhat underdeveloped.

We argue that the primary cause of this underdevelopment lies in methodological limitations, which hinder thorough exploration and understanding. Indeed, tourism scholars tend to rely on traditional research methods such as surveys, interviews and observations when studying animal-based tourism, thus privileging the human over the non-human agent (Bertella, 2022). While more innovative research approaches, including video methods, have started to gain ground in tourism studies, the focus has remained primarily human-centred (Bursta *et al.*, 2023; Masset *et al.*, 2024; Rakić & Chambers, 2010). However, there has been a recent surge of studies calling for and exploring innovative methodological approaches that acknowledge the agency of non-human animals in tourism (e.g. Äijälä, 2021; Bertella, 2022; Haanpää *et al.*, 2019). These studies have primarily drawn on methodological developments from the fields of animal geography (Buller, 2015) and moving image methodologies (Lorimer, 2010; Rokka & Hietanen, 2018; see also Vannini, 2020).

In light of this gap, this chapter focuses on using video as an essential component in both the research process and the dissemination of knowledge about human-animal encounters in tourism. To illustrate our arguments, we draw upon two different research processes related to human–non-human relations within sled dog kennels and reindeer farms situated in Finnish Lapland. While the tourism industry is full of examples where individuals or packs of animals from various species serve as the primary attraction or play a significant role in the tourism experience, such as wildlife-watching, fishing, working animals, and animals as food (Fennell, 2012; Hoarau-Heemstra & Kline, 2022; Rickly & Kline, 2021), Finnish Lapland offers a unique context for studying human-animal encounters in tourism.

Sled dogs and reindeer are not only popular among tourists, but they also contribute to the brand value of this Fennoscandian region (García-Rosell & Äijälä, 2018). The promotion of Finnish Lapland as a destination thrives within a media landscape dominated by visuals of nature and non-human animals, which are produced by the industry and reproduced by tourists (see Masset *et al.*, 2024). Viewed through this lens, destination marketing appears as a primarily human-centric effort, where animals are confined to a background role, thus neglecting their agency in the co-creation of the tourism experience. Hence, the unproblematised power imbalance between the human and non-human animal gaze is inherent in tourist representations (Äijälä, 2021; Tomassini & Bertella, 2023). These representations renew and affirm the idea of tourism as legitimised interest towards an otherness and processes of othering conceived in terms of species diversities. Hence, the current tourist landscape neglects non-human animals as playing their full part in the co-creation process of tourism experiences.

We believe that the dilemmas related to animals' exclusion and being regarded as 'other' in tourism are spatial, as tourism spaces are still based on and articulated by neoliberal capitalism (Tomassini & Lamond, 2023). Hence, we move beyond these kinds of articulations and start from the premise that space is a constantly evolving material and social phenomenon, where 'social' is not purely a human domain (Massey, 2005; see Ingold, 2013). Animals and non-human subjects, which are valued by tourists and tourism organisations for their otherness, contribute to the relationality of tourism spaces (Dashper, 2020). In much the same way, as argued by Dashper (2020), the act of 'othering' non-human animals takes on a political dimension, influencing our behaviour towards these creatures and enabling us to justify care, but also mistreatment and even abuse.

In this chapter, we explore the possibilities of video methods as a means to co-create more inclusive spaces of tourism that make space for both nonhuman others and otherness. Firstly, we examine how video supports data collection and analysis that enables us to access and comprehend the constantly unfolding intimate and intangible spaces and times of touristic mushing (i.e. dog sledding for commercial purposes) (Äijälä, 2022; see Brown & Banks, 2015). Secondly, we explore the role of videography in evoking the relational and affective qualities of spaces of human-animal encounters in the context of dog sledding tours and reindeer sleigh rides (see Haanpää *et al.*, 2019). Thirdly, we discuss how video methods open up new avenues for co-creating space for non-human animals in tourism futures.

In doing so, we respond to the call for new methodological approaches to create means to bring new ways to acknowledge non-human animals in co-creation processes through versatility-embracing tourism knowledge instead of human-centrism (see Lorimer, 2010; Tomassini & Bertella, 2024). The co-creational approach could contribute to transgressing the still-reigning conceptual and material spaces of tourism based on the ideal Western model, and develop more inclusive spaces for non-human animals in the tourism futures. More inclusive spaces for all agents will most likely not mean the non-use of animals in tourism, at least in the near future.

However, as Fennell (2022) argues, they could, lead to a stronger agency and voice for animals participating in particular activities and attractions. To envision such futures, we draw on Bergman *et al.*'s (2010) typologies of future studies, which help in identifying a utopian scenario in tourism development. Additionally, Inayatullah's (1998) critical futures research approach provides a framework to critically explore the dominant narratives and underlying structures that shape human-nonhuman relations in tourism. By incorporating these perspectives, we open possibilities for reimagining multispecies tourism spaces where non-human animals are not merely passive participants but active co-creators of the tourism experience.

Non-Human Animals as Co-Creators of Tourism Space

The asymmetric power relation between animals and humans working in tourism is evident. As Hoarau-Heemstra and Kline (2022) argue, ethical dilemmas relative to animal exploitation and abuse, environmental degradation, social appropriation, commercialism and economic greed are never far away from human-animal encounters in tourism. These dilemmas may relate to the contractarian thinking of tourism operators and ontologically and epistemically constructed barriers by theorists (Fennell, 2022). We believe these dilemmas and barriers are spatial, as non-human animals are much too often excluded from the polished but highly contested tourism spaces (Tomassini & Lamond, 2023). Indeed, human-centric power relations play a critical role in placing boundaries on the extent and type of non-human otherness deemed acceptable within a tourism space (Dashper, 2020).

In the tourism context, the policies and practices of the tourism industry frame the encounters significantly and universally, while conditions such as the weather and character of the animals, guides and tourists shape the encounters at a local level. Hence, our endeavour is based on the premise that animal-human encounters cannot take place without a backdrop, namely the space, which is fundamental to this relationship. Such an approach is grounded in Massey's (2005) conceptualisation of space as a lively, heterogeneous, and political response to global bio-capitalism, and the social inequalities generated by unbalanced power relations, where 'social' is not purely a human domain (see Ingold, 2013).

By following this path, we go beyond the presumption that the relationship between non-human animals and humans is a dualistic one. Instead, tourism space and the way human and non-human beings create and relate to it is a trinary rather than a dualistic relation. What follows is that tourism activities are shaped to a large extent through doing by both animals and humans, as they act and react to the conditions surrounding them (Äijälä, 2022; see Gray, 2014). Non-human animals play their full part in the construction and development of space.

To understand and analyse tourism space, we draw on the ideas of co-creation of value, which has been an influential conceptual discussion in theorising on tourism services and their execution (see e.g. Campos *et al.*, 2018; Bertella, 2014; Haanpää *et al.*, 2016; Lugosi, 2018). In co-creation discussions, the understanding of value has been rather human-centred (e.g. Ribeiro *et al.*, 2023; cf. Bertella, 2014). In the tourism context, value can be understood as value-in-experience that is intra- and inter-subjective, both lived and imaginary, temporal in nature, and emerging from individually determined social contexts (e.g. Helkkula *et al.*, 2012). To co-create value, the different agents need to mobilise their knowledge and skills. Based on this idea, co-creation is at its core an activity based on the place and everyday practices of tourism firms (García-Rosell *et al.*, 2019). Following Lugosi (2014), we understand co-creation as 'a dynamic process in which multiple objects, actions and agencies interact' (2014: 177).

To comprehend co-creation, we focus on agency and take it as essentially relational (Äijälä, 2021; Bertella, 2014; Fennell, 2022). Agents, whether human or animal, are involved within particular social and cultural relations, and they can modify their behaviour and 'voice' (Fennell, 2022) within those relations according to changing circumstances. Agents are distinguished regarding their place within the multidirectional social relations taking place in a particular environment (Carter & Charles, 2013). Another feature of agency is skill, which is required in acting and taking one's place in a particular environment. Skilled action does not come ready-made, but develops through biologically and socially inhabiting the place (Ingold, 2011; Tsing, 2014). Tourism activities require the skilled action of both animals and humans.

It would still be misguided to claim that non-human animals and humans have mutual and symmetrical capabilities to become skilled agents in the tourism industry, and thus hold equal opportunities, or 'voice', to consent or deny consent through their actions (see Fennell, 2022). There are instances where the tourism industry is bent on profit and pleasure at the expense of animals (e.g. Fennell & Sheppard, 2011). Still, animals are not just passive participants, as they shape tourism practices through their actions and reactions and bring their own experiences and life history to the human-animal encounter (Notzke, 2019). Hence, we do not take agency as a universal property and equal ability of all humans and particularly non-human animals working in the tourism industry to effect change in their own lives and the industry. Our stance is rather based on a belief in the possible emergence of shared knowledge, practices, and ultimately culture (Buller, 2012).

To understand the knowledge at play in co-creation processes inclusively, we read knowing as embodied movement is connected to the spaces where it takes place (Haanpää, 2022). Thus, we subscribe to processual and relational onto-epistemologies understanding the tourism spaces as networked entities, consisting of moving bodies navigating with/in them (see Tomassini & Lamond, 2023; Markuksela & Valtonen, 2019). This calls for recognising and acknowledging the capabilities and possibilities of participation of different agents in co-creation processes. To consider the co-creation processes, we, instead of universalism, take a more place-based approach and focus on particular animals working in particular tourism environments. Following Bertella (2014), we include non-human animals as active and central co-creators in the local tourism experience network. The quality of experience is based on a shared set of knowledge and practices between human and non-human animals. Since knowledge is a central resource for building value in these processes, understanding and theorising on different forms of knowledge calls for novel methodologies (e.g. Haanpää, 2022). We consider how video methods can help us to understand the co-creation of value between and for different agents in the process.

Video Methods as a Means to Shed Light on the Co-Creation of Tourism Spaces

Here we advance an argument about the methodological procedures, practical techniques and epistemological underpinnings of mobile video ethnography and videography when applied in the context of animal-based tourism. The first section refers to the first author's experiences during the fieldwork of his PhD project between summer 2019 and spring 2020 in a sled dog kennel in Finnish Lapland. The project aimed to bring to light the living and moving sled dog-human-place relations in tourism (Äijälä, 2022). When referring to this project the expression 'first author' is used.

The second section refers to a videographic study made in a research project focusing on animal welfare in tourism. The project, implemented from 2016 to 2018, aimed to promote the welfare of animals used in tourism services in Lapland, as well as to provide opportunities for animal-based tourism firms and other stakeholders to benefit from responsible practices in the industry. While the project initially focused on animal welfare as a means to meet human needs, it was the videography that directed our attention to approach animal-based tourism from a nonhuman perspective. All three authors of this chapter were involved in the project and are referred to as 'the authors'. Outside this subchapter, the personal pronoun 'we' refers to the joint process of reflections and discussions between the three authors of this book chapter.

Mobile Video Ethnography: Gaining Access to the Intimate Spaces of Human-Animal Encounters in Tourism

During his ethnographic fieldwork, the first author was struggling to keep up with all the intense actions included in everyday life at the local sled dog kennel in Finnish Lapland. He felt that 'traditional' ethnographic techniques – based mainly on traditional participant observation and talk – were insufficient in accessing the intimate and intangible spaces of human-sled dog encounters. Even the use of a fixed-position or a handheld camera – not to mention interviewing only the human participants – would have made it possible to keep up with and be involved in the intensity of touristic mushing (Äijälä, 2021; 2022: see Brown & Banks, 2015). He had conducted some initial experiments with wearable action cameras, but after realising the constraints of more traditional research techniques, he started to delve deeper into the possibilities offered by the minicam to apply a method that could 'go along' (Brown & Banks, 2015; Vannini, 2017) as the humans and dogs engaged in their daily life at the kennel.

As the first author failed to enter the intimate spaces of human-dog encounters by becoming a more-than-tourist musher, he turned to a research agenda of exploring life at the kennel with the minicam video, which is at the heart of mobile video ethnography involving the use of video as a 'go along' method to follow the research participants' ordinary mobile routines and observing how they engage with and reflect on their daily life. It is a means of gaining access and creating an understanding of social worlds emerging through the interactions of living agents (Äijälä, 2022; see Vannini, 2017). In practice, three separate GoPro devices were employed: a Dog GoPro, a Musher GoPro and a Researcher GoPro, and the video material was complemented with action camera material filmed by one of the guides. As described in the vignette at the beginning of this chapter (see Figure 9.1), the minicam video allowed access to the intimate spaces of touristic mushing.

The action camera footage made little sense by itself, with the first author's limited experience of dog sledding. It was wide open to missing aspects or even misunderstandings of what the participants were doing, thinking and feeling. Thus, the footage cannot be taken as upfront evidence of reality. Rather, the footage was a trace that animates the dog-human-place trinary for which we have little established vocabulary (Brown & Banks, 2015; Bursta *et al.*, 2023). For Lorimer (2010), the footage helps to evoke novel spaces for performance and participatory analysis. The participatory analysis took place through video review, or audiencing, a process in which each individual trace enabled a process of reflection, discussion and understanding by the human informants, namely the two mushers who carried the Musher GoPro. Johnston (2008) calls this tactic 'responsible anthropomorphism', wherein those who have long-term, day-to-day contact with animals are allowed to speak for them. It could also be argued that the dogs were involved in this participation as they had a chance to articulate their stances through the video footage (Äijälä, 2022). Through the fieldwork the mushers and partially the dogs were able to articulate meaning for themselves. The video footage and the accompanying audiencing process created rich material for a story that emerged from engagement with the meaningful micro-spaces and micro-events in an overwhelmingly packed and disordered reality.

Videography: Relational and Affective Qualities of Videos to Unlock New Ways of Co-Creating Tourism Spaces

In a research project concerning animal welfare in tourism, we conducted videographic research on the topic. This was the very first research project to study animal-based tourism services in Finnish Lapland, and the videographic research originally examined consumer values in relation to the use of animals in tourism (Haanpää *et al.*, 2021). Videography as a method has its underpinnings in the ethnographic tradition, but besides observing and collecting material through video, the aim is also to disseminate the knowledge gained with a video or moving visuals (Petr *et al.*, 2015; Seregina, 2018). In this chapter, we focus especially on the last part of the videographic research process: the video presentation and engaging audiences with it. We examine how the editing of video material and the film screenings contribute to creating new ways of seeing and understanding tourism spaces (see Haanpää *et al.*, 2019; Rokka & Hietanen, 2018). We argue, in line with Rokka and Hietanen (2018), that this is especially due to the relational, affective and performative qualities of videographic research.

Given that our interest was first in consumer values, we carried out filmed interviews with international tourists. Besides the interview material, we filmed and observed animal-based tourism activities in Lappish tourism companies with and without the interviewees. In totality, almost seven hours of video material was obtained during the course of research (Haanpää *et al.*, 2019). The process of watching and editing the filmed material began before the whole course of filming was completed. This gave us access back to the field and allowed us to contemplate the events and moments captured but also felt. However, watching the filmed material also produced new feelings and affectivities and it was through the editing process where we noticed the powerful presence, or the agency, of animals (Haanpää *et al.*, 2019: 103). This led to us crafting a whole new videography project where we examined human-animal relations in tourism as multi-species assemblage (Haanpää *et al.*, 2019). Besides the videography, we also made the videography according to our original plans, illustrating the different ethical stances tourists can have towards animals.

The two videographies crafted were titled 'Who is ethical tourist?' and 'The disruptive "other"?' They were ten and five minutes in length. 'Who is ethical tourist?' was aimed at the general public and industry actors to provoke discussion about tourists' values towards animals, and it was only produced in video format for screening purposes. 'The disruptive "other"?', exploring human-animal relations in tourism as multispecies assemblage, was more aimed at academic audiences to contemplate the phenomenon, and it was also published as an academic contribution alongside the screenings (see Haanpää *et al.*, 2019). We screened both videographies on different occasions. 'The disruptive "other"?' was screened at two academic conferences: the '2nd Peaceful Coexistence Colloquium – Reimagining Ethics and Politics of Space for the Anthropocene' (2017) and the 'Association for Consumer Research Conference' (2020), and also at a smaller symposium called the 'Living Ethics Seminar' (2017). Again, 'Who is the ethical tourist?' was screened at the Lapland Tourism Parliament (2017), which is an annual, regional industry event, and at the Matka Travel Fair (2018), the biggest annual, national industry event in Finland (see Figure 9.2). Both videographies were also used and screened several times in the classroom as part of our teaching.

As described in the vignette at the beginning of this chapter (see Figure 9.2), the videographies almost always generated conversation and the more-than-representational qualities of video brought the nonhumans and humans alive in these spaces in a way that more traditional ways of representation probably could not have done. The potentiality of video material to intensify the affective qualities of lived events was present in these situations and thus video provided space for reflection, learning and means of communication (Rokka & Hietanen, 2018; Seregina, 2018; Vannini, 2015). The conversations evoked by the videographies also made us realise and appreciate these qualities on a more profound level. For example, we first had quite extensive textual passages in 'The disruptive "other"?' that were questioned in one of the academic screenings, in that they distracted the viewing experience away from the nonhuman subjects. We then edited most of the text out and let the moving image affect the viewers' interpretations and feelings. Following Rokka (2022), video methods hold the ability to shock our habituated ways of thinking about tourism spaces and cast a novel light on tourism futures in which nonhuman animals play a more inclusive role.

Co-Creating More Inclusive Spaces in Tourism Futures Through Video Methods

In the field of tourism studies, the potential of video has been underutilised, while the current research has been more *with* video format (e.g. Bursta *et al.*, 2023). Then the footage works as a means to collaborate with the research participants to create evocative and affective spaces and events of the human-animal encounters that are meaningful to research participants (Brown & Banks, 2015). Although mobile video ethnography has great potential to 'go with' the research subjects, including non-human animals, and follow the action and flow of unfolding micro and macro movements and spaces, the endeavours to date have somewhat failed to disseminate these spaces, that is, research insights, to wider audiences (see Vannini, 2015).

Even though there are differences between the two methods we have discussed here, the movement and affective qualities of video connect them. The essence of video methods lies in the way they create the possibility of accessing the intimate, intangible, multiple and even chaotic spaces and events of human-animal encounters in tourism. The video footage of going along with the research participants and the subsequent audiencing process evoked their memories and feelings about their actions to add another layer of sense-making and knowledge about the co-creation process of tourism spaces (Äijälä, 2021; see Brown & Banks, 2015). Within videography, even larger audiences could participate in the co-creation process (Haanpää *et al.*, 2019). As Vannini and Stewart (2017) suggest, the video footage works as an audio-visual tool for documenting and communicating the spaces and events for the benefit of distant audiences.

In our research endeavours, we have witnessed video methods co-creating space for nonhumans in tourism in different contexts. Through close collaboration with the animal caretakers, video methods can bring new managerial implications as the staff might be willing to change their own, perhaps unconscious and incorrect, practices when seen through the video trace (Äijälä, 2022). In the research context, the videographic method led us to acknowledge the nonhuman space in our own research through its affective qualities. The screenings of our research findings as videographies for different audiences, industry, students and fellow academics, have opened up space for nonhumans more effectively than only verbal or textual presentations would have (Haanpää *et al.*, 2019).

Although these methods bring animal agency to the fore and perhaps even give a better 'voice' for animals, we have to be cautious about the unequal relation of power between the researcher, human and animal research participants. The power imbalance is addressed directly in the post-production phase, including audiencing, editing and broadcasting, which mobilise the video footage to further layers of sense-making and knowledge (Brown & Banks, 2015; Haanpää *et al.*, 2019). The nonhuman animals are unable to participate in this process as we do not have a shared oral language, which makes it difficult, perhaps impossible, for them to have control over how their actions are monitored, recorded and made visible (Brown & Banks, 2015; Äijälä, 2022).

In our work we have concentrated on the co-creation of tourism spaces with nonhuman animals that provide an accessible, popular and photogenic case study. Undoubtedly, our work outlined here favours animals whose size and modes of relating to their environment permit proximate, mostly safe and long-term relations (Lorimer, 2010). However, tourism scholars have acknowledged that the presence of innumerable forms of nonhuman animals within the spectrum, from unpleasant head lice

(Benali & Ren, 2019) to charismatic whales (Mallard, 2019), matters for tourism. With video-based methods we have outlined here that it might be difficult to make available for reflection the intimate tourism spaces that some nonhuman animals, for example head lice, co-create as tourism spaces. Hence, video-based methods do not offer an all-encompassing solution to overcome the issue of how we value certain charismatic animals over many others.

We still believe that when used sensitively with a decent amount of skill, video methods bring us closer to acknowledging the power imbalance, as they help us in thinking about animals not as mere objects, but as agents that shape our ways of thinking about the world and ourselves. If non-human animals are agents that intentionally or unintentionally 'gesture towards the future' (Tsing, 2014: 28) making new worlds, then these video methods could be a gesture towards the future of going along and better understanding and disseminating these world-making processes. Following Vannini (2015), the role of video should not be viewed as a reproductive one but as a poetic one, whose essence lies in the way it 'makes knowledge' and facilitates the practical and affective dimensions of nascent social reality itself.

Video as a relational 'media practice' is embedded in and produces social relations and is thus one that is capable of inducing change (Hietanen *et al.*, 2014). Hence, it is not merely a description of past events but holds the potential to help us imagine new ways of being, acting and relating in the world (Haanpää *et al.*, 2019; Hietanen & Andéhn, 2018). This is underlined in videographic research, where editing the final presentation is always a relational and political act (Rokka & Hietanen, 2018; Vannini, 2015). For Vannini (2015), the making of choices in the production and postproduction phase is an aesthetic, political and ethical process focused on constructing knowledge (see Haanpää *et al.*, 2019). For example, we as researchers have to choose how to deal with the situation if the video footage includes unethical practices related to animal keeping.

Through producing social relations also with distant audiences, video methods hold great potential in co-creating more inclusive human-animal spaces in tourism futures. They shed light on the fundamental role of lively nonhuman animals in the tourism industry and bring to the fore the central part played by non-human animals in shaping and negotiating tourism spaces (Äijälä, 2022). Following Ingold (2013), they are not about looking at animals as though they were humans, or at humans as though they were animals, but rather about looking for ways of constituting social life that allow for their differences – to be differently intelligent.

Engagement in video-based methods could provide alternative shifts in perspective, even 'shock to thought' (Rokka, 2022), for tourism scholars, educators and practitioners. As Rokka (2022) argues, video as a reflexive tool could be used to generate trust, transparency, and facilitate communication between the researcher and the informants. This 'collaborative videography' (Rokka, 2022) could help us to transgress the barriers standing in the way of a stronger voice for non-human animals in tourism (Fennell, 2022). Thus, video as an important reflexive tool contributes to overcoming the processes of othering and creating more inclusive spaces that accommodate both nonhuman others and otherness in the human-centric tourism sphere.

The more inclusive multispecies spaces created through collaborative videography have implications for tourism futures. Most likely, the shock to thought and providing a stronger voice for non-human animals in tourism will not lead to non-use or even collapse of animal-based tourism, at least in the near future. A more precautionary and realistic option is that these methods lead tourism agents to foster reconciliation with all to whom we owe care (Higgins-Desbiolles, 2024; see also Tallberg *et al.*, 2021). Following Fennell (2022), these methods contribute to stronger animal-informed consent in tourism and, in turn, evolve into a long-term understanding of the suitability of animals for use in particular activities and attractions in particular spaces. Tourism education, research and practice just has to be open to novel approaches and face the shock to thought to secure better tourism futures for all.

Conclusions: Co-Creating Utopian Tourism Futures Through Videography

There is a growing interest among tourism scholars and practitioners in better acknowledging the role of non-human animals in their field. While theoretical advancements in this domain continue to evolve, methodologies for studying these interactions remain underdeveloped (Äijälä, 2021). In this chapter, we have explored the potential of moving image methodologies (e.g. Lorimer, 2010) to evoke human-animal-place interactions in tourism. We have theorised human-animal-space relations while also demonstrating the capacity of mobile video ethnography and videography to enhance understanding and dissemination of knowledge about these encounters. By applying Bergman *et al.*'s (2010) typologies of future studies, we envision utopian futures for tourism in relation to non-human animals.

Utopian futures, inspired by critical futures research (Inayatullah, 1998), envision non-human animals as legitimate co-creators of tourism spaces. In this future, videography and other participatory methodologies will serve as transformational tools, fostering new epistemologies that challenge entrenched speciesist hierarchies. The use of critical futures research approaches (Inayatullah, 1998) would allow scholars and practitioners to deconstruct the deeper ideological and institutional frameworks that sustain anthropocentrism in tourism, paving the way for more inclusive and ethical multispecies interactions. In this utopian vision, video-based methodologies become a means of granting animals a form of voice and presence in tourism research and practice. By capturing animals' behaviours, interactions, and responses to tourism environments, videography offers an alternative mode of knowledge production – one that acknowledges their agency and lived experiences. Through collaborative videography, human and non-human subjects co-create tourism spaces, leading to ethical practices where animals are recognised as active participants rather than passive subjects.

The transformative potential of videography extends beyond academia and industry, reaching broader audiences to stimulate critical reflection on human-animal relations in tourism. By disseminating video narratives that highlight non-human agency, tourism stakeholders – from policymakers and DMO to tourists themselves – can engage with alternative perspectives that challenge exploitative practices. In doing so, videographic storytelling becomes a powerful mechanism for shaping multispecies tourism futures rooted in mutual respect and care. Ultimately, this chapter argues that embracing video methods as tools for co-creation is essential in reimagining tourism futures beyond human-centric paradigms. While the transition toward a truly multispecies-inclusive tourism industry may be gradual, the methodological innovations discussed here lay the groundwork for ethical and participatory approaches that foreground the agency of non-human animals (see Äijälä *et al.*, 2021). If tourism is to move toward a more just and inclusive future, it must embrace transformational methodologies like videography, which not only document but actively contribute to the creation of spaces where all beings – human and non-human alike – can thrive.

References

Äijälä, M. (2021) Mobile video ethnography for evoking animals in tourism. *Annals of Tourism Research* 89, 103203.

Äijälä, M. (2022) *Tourist Landscapes as Multispecies Trails: Storying a Mushing Landscape Through Mobile Video Ethnography*. University of Lapland.

Äijälä, M., Jylkäs, T., Rajab, V. and Vuorikari, T. (2021) Designing future wildlife tourism experiences: On agency in human–sled dog encounters. In G. Bertella (ed.) *Wildlife Tourism Futures: Encounters with Wild, Captive and Artificial Animals* (pp. 126–139). Channel View Publications.

Benali, A. and Ren, C. (2019) Lice work: Non-human trajectories in volunteer tourism. *Tourist Studies* 19 (2), 238–257.

Bergman, A., Karlsson J.C. and Axelsson, J. (2010) Truths claims and explanatory claims – an ontological typology of futures studies. *Futures* 42 (8), 857–865.

Bertella, G. (2014) The co-creation of animal-based tourism experience. *Tourism Recreation Research* 39 (1), 115–125.

Bertella, G. (2022) Including the 'voices' of animals in tourism, hospitality and leisure research. In F. Okumus, M. Rasoolimanesh and S. Jahani (eds) *Advanced Research Methods in Hospitality and Tourism* (pp. 199–211). Emerald.

Brown, K.M. and Banks, E. (2015) Close encounters: Using mobile video ethnography to understand human-animal relations. In C. Bates (ed.) *Video Methods: Social Science Research in Motion* (pp. 95–120). Routledge.

Buller, H. (2012) 'One slash of light, then gone': Animals as movement. *Etudes Rurales* 189, 139–154.
Buller, H. (2015) Animal geographies II: Methods. *Progress in Human Geography* 39 (3), 374–384.
Bursta, B., Kvidal-Røvik, T. and Rantala, O. (2023) Tuning ourselves into place: Enhancing multivocality with video. *Qualitative Research* 0 (0).
Campos, A.C., Mendes, J., do Valle, P.O. and Scott, S. (2018) Co-creation of tourist experiences: A literature review. *Current Issues in Tourism* 21 (4), 369–400.
Carr, N. and Broom, D.M. (eds) (2018) *Tourism and Animal Welfare*. CABI.
Carter, B. and Charles, N. (2013) Animals, agency and resistance. *Journal for the Theory of Social Behaviour* 43 (3), 322–340.
Danby, P., Dashper, K. and Finkel, R. (2019) Multispecies leisure: Human-animal interactions in leisure landscapes. *Leisure Studies* 38 (3), 291–302.
Dashper, K. (2020) More-than-human emotions: Multispecies emotional labour in the tourism industry. *Gender Work Organization* 27 (1), 24–40.
Fennell, D.A. (2012) *Tourism and Animal Ethics*. Routledge.
Fennell, D.A. (2022) Animal-informed consent: Sled dog tours as asymmetric agential events. *Tourism Management* 93, 104584.
Fennell and Sheppard (2011) Another legacy for Canada's 2010 Olympic and Paralympic Winter Games: Applying an ethical lens to the post-games' sled dog cull. *Journal of Ecotourism* 10 (3), 197–213.
García-Rosell, J.-C. and Äijälä, M. (2018) Animal-based tourism in Lapland. In J. Ojuva (ed.) *Animal Welfare in Tourism Services: Examples and Practical Tips for the Well-Being of Animals Used for Tourism in Lapland* (pp. 10–24). Lapland University of Applied Sciences.
Gray, J. (2014) Hefting onto place: Intersecting lives of humans and sheep on Scottish hills landscape. *Anthrozoös* 27 (2), 219–234.
Haanpää, M. (2022) Co-creation as choreography. *Qualitative Market Research* 25 (5), 614–624.
Haanpää, M., García-Rosell, J.C. and Tuulentie, S. (2016) Co-creating places through events: The case of a tourism community event in Finnish Lapland. In A. Jepson and A. Clarke (eds) *Managing and Developing Communities, Festivals and Events* (pp. 34–49). Palgrave Macmillan UK.
Haanpää, M., Salmela, T., García-Rosell, J.-C. and Äijälä, M. (2019) The disruptive 'other'? Exploring human-animal relations in tourism through videography. *Tourism Geographies* 23 (1–2), 97–117.
Helkkula, A., Kelleher, C. and Pihlström, M. (2012) Characterizing value as an experience: Implications for service researchers and managers. *Journal of Service Research* 15 (1), 59–75.
Hietanen, J. and Andéhn, M. (2018) More than meets the eye: Videography and production of desire in semiocapitalism. *Journal of Marketing Management* 34 (5–6), 539–556.
Hietanen, J., Rokka, J. and Schouten, J.W. (2014) Commentary on Schembri and Boyle (2013): From representation towards expression in videographic consumer research. *Journal of Business Research* 67 (9), 2019–2022.
Higgins-Desbiolles, F. (2024) The end of tourism? Contemplations of collapse. *Journal of Tourism Futures* 10 (3), 476–485.
Hoarau-Heemstra, H. and Kline, C. (2022) Making kin and making sense of human-animal relations in tourism. *Ecological Economics* 196.
Inayatullah, S. (1998) Causal layered analysis: Poststructuralism as method. *Futures* 30 (8), 815–829.
Ingold, T. (2011) *Being Alive: Essays on Movement, Knowledge and Skill*. Routledge.
Ingold, T. (2013) Anthropology beyond humanity. *Suomen Antropologi: Journal of the Finnish Anthropological Society* 38 (3), 5–23.

Johnston, K. (2008) Beyond the clearing: Towards a dwelt animal geography. *Progress in Human Geography* 32, 633–649.

Lorimer, J. (2010) Moving image methodologies for more-than-human geographies. *Cultural Geographies* 17 (2), 237–258.

Lugosi, P. (2014) Mobilising identity and culture in experience co-creation and venue operation. *Tourism Management* 40, 165–179.

Lugosi, P. (2018) Disruptive ethnography and knowledge co-creation. In C. Ren, G. Jóhannesson and R. van der Duim (eds) *Co-Creating Tourism Research* (pp. 55–72). Routledge.

Mallard, G. (2019) Regulating whale watching: A common agency analysis. *Annals of Tourism Research* 76, 191–199.

Markuksela, V. and Valtonen, A. (2019) Dance with a fish? Sensory human-nonhuman encounters in the waterscape of match fishing. *Leisure Studies* 38 (3), 353–366. https://doi.org/10.1080/02614367

Masset, J., Decrop, A. and Frochot, I. (2024) Videography in tourism research: An analytical review. *Tourism Management* 24 (102), 104869.

Massey, D. (2005) *For Space*. SAGE

Notzke, C. (2019) Equestrian tourism: Animal agency observed. *Current Issues in Tourism* 22 (8), 948–966.

Petr, C., Belk, R.W. and Decrop, A. (2015) Videography in marketing research. Mixing art and science. *Arts and the Market* 5 (1), 73–102.

Rakić, T. and Chambers, D. (2010) Innovative techniques in tourism research: An exploration of visual methods and academic filmmaking. *International Journal of Tourism Research* 12 (4), 379–389.

Ribeiro, T.D.L.S.A., Costa, B.K., Ferreira, M.P. and de Lamônica Freire, O.B. (2023) Value co-creation in tourism and hospitality: A systematic literature review. *European Management Journal* 41 (6), 985–999.

Rickly, J.M. and Kline, C. (eds) (2021) *Exploring Non-Human Work in Tourism. From Beasts of Burden to Animal Ambassadors*. De Gruyter Oldenbourg.

Rokka, J. (2022) Videography as a 'Shock to Thought'. *Journal of Customer Behaviour* 21 (1–2), 26–35.

Rokka, J. and Hietanen, J. (2018) On positioning videography as a tool for theorizing. *Recherche et Applications en Marketing (English Edition)* 33 (3), 106–121.

Seregina, A. (2018) Engaging the audience through videography as performance. *Journal of Marketing Management* 34 (5–6), 518–535.

Tallberg, L., García-Rosell, J.C. and Haanpää, M. (2022) Human–animal relations in business and society: Advancing the feminist interpretation of stakeholder theory. *Journal of Business Ethics* 180 (1), 1–16.

Tomassini, L. and Bertella, G. (2024) The human gaze at animals and the missing animal gaze in tourism studies. *Current Issues in Tourism* 27 (3), 350–355.

Tomassini, L. and Lamond, I. (2023) Rethinking the space of tourism, its power-geometries, and spatial justice. *Journal of Sustainable Tourism* 31 (12), 2825–2838.

Tsing, A. (2014) More-than-human sociality: A call for critical description. In K. Hastrup (ed.) *Anthropology and Nature* (pp. 27–42). Routledge.

Vannini, P. (2015) Afterword. Video methods beyond representation: Experimenting with multimodal, sensuous, affective intensities in the 21st century. In C. Bates (ed.) *Video Methods. Social Science Research in Motion* (pp. 230–240). Routledge.

Vannini, P. (2017) Low and slow: Notes on the production and distribution of a mobile video ethnography. *Mobilities* 12 (1), 155–166.

Vannini, P. (2020) Introduction. In P. Vannini (ed.) *The Routledge International Handbook of Ethnographic Film and Video* (pp. 1–11). Routledge.

Vannini, P. and Stewart, L.M. (2017) The GoPro gaze. *Cultural Geographies* 24 (1), 149–155. https://doi.org/10.1177/1474474016647369

10 Exploring the Twilight Zone: Anthropology, Science Fiction and Tourism Futures in Urban Environments

Francesco Vietti

Prologue – October 2022, Turin, Italy

The Space Pirate Captain Harlock statue in front of the MUFANT, Turin
© Francesco Vietti

Welcome to our exploration of the urban environment of the northern suburbs of Turin, the capital city of Piedmont region in Italy!

We are in front of the entrance of the MUFANT, the Museo del Fantastico e della Fantascienza (in English, Museum of Fantasy and Science Fiction), right next to one of the sculptures of the Park of the Fantastic created by the museum in this part of the outskirts of the city. You would have certainly recognized the unmistakable profile of Captain Harlock, the space pirate born from the imagination of Japanese manga artist Leiji Matsumoto in 1977. Today, Captain Harlock can certainly be a good travel companion on our tour 'in the twilight zone'. Firstly, because he is a great space explorer, always looking for new adventures in the cosmos aboard his spaceship, Arcadia. Secondly, because this character invites us to reflect on the ethical and political dimensions of the journey. In fact, Matsumoto's manga tells us about a future Earth, the year is 2977, in which humanity has exploited our world to the point of being forced to plunder the resources of other planets to survive. This is an apathetic, indifferent humanity, from which Capital Harlock stands out for his thirst for justice, his unwavering will to fight against all forms of inequality and oppression (McCarthy & Ashmore, 2021). Precisely for this reason, I find the sentence that MUFANT decided to write on the pedestal of the statue very significant. This is the famous quote from Martin Luther King: No one is free until we are all free.

Let us now look at the sculpture behind you: Frankenstein's profile accompanied by the well-known statement attributed to Franco Basaglia, the Italian psychiatrist who, in the 1970s, revolutionised mental health practices by fighting for the rights and dignity of people with mental fragility: Seen up close, no one is normal. I would like us to take this reference to seeing, to looking at something closely, as a tool to get us started with our urban exploration. Anthropology and science fiction invite us to observe the city carefully, to reflect on the interstices, to deeply exercise our ability to recognise what we have too often rather learned to unsee. I take this concept from a beautiful novel by China Miéville, a writer who we can imagine as a guide on this journey of ours. Miéville is a British author, born in 1972. He is one of the most significant contemporary writers of science fiction and so-called 'urban fantasy'. But he is also an anthropologist by training, with studies in social anthropology at Cambridge. In his novel The City and the City Miéville (2009) tells us about two cities or rather a city divided into two parts, Beszél and Ul Qoma. Beware, however, it is not a city separated by a wall or a border. Beszél and Ul Qoma are two overlapping cities, occupying the same space. The inhabitants of one and the other live side by side, but from an early age they are taught to unsee each other, to look away from each other.

Today, what I would like to suggest, is that Beszél and Ul Qoma also talk to us about what happens every day in Turin and in any other city at global level. We, too, learn to unsee the others. Many residents of the city, human and other-than-human, are invisible to our eyes. Let us, therefore, try to think of anthropology and science fiction as the lenses of a pair of glasses that we can put on to broaden our perspective of the city and its human and non-human inhabitants. In this regard, I would like to read to you a brief passage of the book where Miéville evokes the very idea of seeing the city through different eyes, finding ourselves to be tourists, marvelling travellers passing through an urban space that we thought of as familiar, and which instead we suddenly discover to be something else, different, we could say alien:

> But pass through Copula Hall and she or he might leave Beszél, and at the end of the hall come back to exactly (corporeally) where they had just been, but in another country, a tourist, a marvelling visitor, to a street that shared the latitude-longitude of their own address, a street they had never visited before, whose architecture they had always unseen, to the Ul Qoman house sitting next to and a whole city away from their own building, unvisible there now they had come through, all the way across the Breach, back home. (Miéville, 2009: 70)

Miéville calls places like Copula Hall crosshatch. As a kind of twilight zone, through these areas it is possible pass from a city to another. This seems to me a powerful image capable of evoking one of the possible futures of urban tourism. A future in which we will be able to breach the (un)visible borders that segregate our cities. A future in which the others will not be mere objects of our processes of othering but subjects capable of transforming through their otherness our own identity as denizens of the city.

Well, now we are ready to go!

The Anthropologist's Guide to the Galaxy

Anthropology and science fiction are united by the epistemological centrality covered by the theme of travel (Stover, 1973; Collins, 2021; Dei *et al.*, 2023). It is precisely the journey, in fact, that enables that 'cognitive estrangement', foundational to both fields, making familiar what is not, and simultaneously observing with a 'look from afar' what is usual. Perhaps, the best known 'common ancestor' shared by anthropology and science fiction in their cultural criticism is the interplanetary journey of Voltaire's *Micromégas* (1752). The Enlightenment approach is evident in the effort to reflect on society as if looking at it through a mirror, viewing the Earth as another world. Inhabitants of Sirius and Saturn (two 'space tourists' ante litteram) philosophize about its strangeness. The meaning of the anthropological journey has been extensively discussed, within the broader history of the scientific journey, and in the perspective summarised by 'long route' formula. As eloquently formulated by Remotti

(1990), anthropological research avoids any 'shortcut' to theorize about humans, choosing the 'long route' through otherness, the plurality of differences, and the challenge of the elsewhere. However long, the journey always has as its final destination, as the return home. This 'here' that will appear in a different light to the researchers, and that they can grasp because they have been 'there'.

In this chapter, I would like to reflect on the futures of tourism by interweaving anthropology and science fiction. The starting point of my analysis is the role that the process of othering and the encounter with the other have played in the history of these two fields. Anthropology and science fiction emerged as fields of human knowledge and creativity at the same time and place of origin. In the last decades of the 19th century, in Victorian London, anthropology became an academically recognised discipline, and the first science fiction novels gained a wide readership. For both early anthropologists and early science fiction writers, the experience of otherness came as result of journeys through space and time (Vietti, 2023).

The Anthropological Expedition to Torres Straits, organized by the University of Cambridge in 1898 between Australia and Papua New Guinea marks a fundamental milestone in the affirmation of the ethnographic method of field research and the first completed attempt at a holistic study of the relationship between humans and the environment in the Pacific area. In the same year, the first edition of *The War of the Worlds*, by Herbert George Wells (1898), was published. The London writer, rightly considered the father of science fiction, here stages the first contact between aliens and terrestrials, representing it in the dramatic tones of invasion and the overthrow of the stronger over the weaker. In describing the destruction of London by ruthless and technologically advanced aliens, the novel's not-too-veiled intent is to force its readers to put themselves in the shoes of the people victimized by colonization (Wells, 1898). Also, in that last decade of the 19th century, while evolutionary anthropologists collected and compared myths and traces of the 'primitive people' of the globe to line up the stages of human civilization, Wells (1895) printed *The Time Machine*. In this science fiction masterpiece, time travel pushes into the future, reaching the world of the year 802701, and even beyond. In Wells, the kinetic force of evolution does not therefore stop at the British society of that time but continues unstoppably towards a future that is far from radiant, in an aberrant involution that leads humanity to divide itself into two distinct races: the Eloj, whose life is spent between amusements and pleasures, and the Morlocks, with a monstrous appearance, confined in tunnels underground. This is a terrible injustice that, as Wells himself notes, was already affirming itself in the London suburbs of his time (Wells, 1895).

During the 20th century, while anthropologists went to every corner of the world to carry out their research, travel became one of the main leitmotifs of science fiction literature. This theme developed in an infinity of variations echoing the two archetypes outlined by Wells: space travel manifested through human explorations of the cosmos and the visits of aliens to our planet, and time travel that also unfolded in dual directions, towards both the past and the future. Countless examples can be brought for each of these forms of science fiction travel, but – perhaps – the title that has had the most lasting impact on the collective imagination is *2001: A Space Odyssey*, a masterpiece that came out both as a novel, signed by Arthur C. Clarke (1968), and as a film directed by Stanley Kubrick (1968), in the same year. This 21st-century Odyssey admirably interweaves the coordinates of time and space, offering us a complex exploration of the relationship between man, the cosmos, life, consciousness, and death. The voyage of the spaceship Discovery is therefore first and foremost an existential exploration, not only towards the most remote universe, but also to the most unfathomable depths of the human soul.

In the 1970s, that characteristic playful, ironic, paradoxical and irreverent spirit, so characteristic of science fiction, made its way around the theme of space-time travel. In an era in which, in this world of ours, the romantic journey of exploration had now definitively left room for the comfort and boredom of mass tourism, tourists made their appearance even in the cosmos, with their equipment of superficiality, taste for fun, awkwardness and desire to see the whole universe in the shortest possible time (Leed, 1991). An icon of this strand of humorous science fiction is certainly *The Hitchhiker's Guide to the Galaxy*, a novel published in 1979 by Douglas Adams, as an adaptation of a radio series written by Adams himself and broadcast on the BBC the previous year. The protagonist of the novel, Arthur Dent, finds himself having to leave Earth, destined to be demolished by the Vogons to make room for a hyperspace expressway. Taking a ride aboard a spaceship equipped with an Infinite Improbability Drive capable of instantly projecting it to every part of the universe, Arthur will be involved in a kaleidoscope of adventures on various planets, which he will gradually discover through the information on a galactic guide (Adams, 1979). It is certainly no coincidence that Adams' book came out a few years after the very first guidebook written by Tony Wheeler (1973), *Across Asia on the Cheap*, which would have given rise to Lonely Planet, the publishing house that is now the world leader in travel guides. The publication of the novel takes place practically simultaneously with that of *Hosts and Guests*, the book by Valene Smith (1978) which marks a crucial moment in the affirmation of the anthropology of tourism.

The anthropological interest in tourism is part of what is commonly referred to as the 'mobility turn' of the social sciences (Urry, 2000). Between the 20th and 21st centuries, mobility has become the prism through which to study the processes of globalisation, the sociocultural transformations in the era of advanced capitalism. The world is characterised by the increase in the possibilities of movement, but also by the growth of global inequalities and forms of (im)mobility control. Drawing on Zygmunt Bauman's formulation (1998), 'tourists' – who travel because they want to – and 'vagabonds' – who travel because they have no other choice – are symbols of the growing divergence between the different 'mobility regimes' of our contemporaneity (Shamir, 2005, Salazar *et al.*, 2014). In this context, cities have emerged as the most relevant context for the ethnographic study of the intersection between forms of mobility and the role that global cultural flows play, today, in the construction of locality (Appadurai, 1996).

Once again, social sciences and science fiction find a meeting point here. Since the beginning of the 20th century, in fact, the city has been the main context where science fiction narratives, through the metaphor of the future, have attempted to highlight the hopes, fears, and contradictions of the present. Just think of *Metropolis*, the novel by Thea von Harbou (1925), which became the screenplay for the masterpiece of expressionist cinema by her husband, the Viennese-born director Fritz Lang (1927). *Metropolis* is projected in the upcoming year 2026 and depicts a city in which social classes live segregated in different layers of urban space. As such, a vision not too dissimilar from that outlined in the same years by the Chicago School of Urban Sociology, inaugurating a tradition of studies dedicated to urban ecology, areas, borders, and ghettos of large industrial metropolises (Park & Burgess, 1925).

A century later, the cities of both anthropology and science fiction are global cities, crossed by a multiplicity of languages and cultures, by flows of moving populations, migrants, refugees, and tourists. Among these is the Hong Kong of the anthropologist Gordon Mathews. With his ethnography *A Ghetto at the Center of the World: Chungking Mansions* (Mathews, 2011), he offers us an insight of a gigantic building in which people from all over the world implement a kind of globalisation from below that makes cities hubs of the mobility of people, goods, money and ideas. But also, Nnedi Okorafor's novel *Lagoon* (2014) is noteworthy for its portrayal of an Afrofuturist Lagos merging sci-fi lagoons, ancient legends, and visions of Nigeria's future. Similarly, Hao Jingfang's novel *Folding Beijing* (2015) describes a dystopic Beijing that revolves around itself. It composes and recomposes like a Rubik's cube, giving substance to the fractures that through the impetuous march towards the future of Chinese society. Even Sam J. Miller's novel *Blackfish City* (2018) imagines a remote floating city in the Atlantic Ocean. The artic melting-pot of Qaanaaq is capable of prefiguring the extreme landing for climate refugees fleeing global warming.

And here we are back to the starting point: the outskirts of Turin, the MUFANT collections and the work of China Miéville. In addition to the novel *The City & the City* (Miéville, 2009), Miéville has published numerous other works in which anthropology and science fiction intertwine in interesting ways, not infrequently in peculiar urban environments. In *Embassytown* (Miéville, 2011) the British writer takes us to a frontier city, on the distant planet Arieka, where human settlers face all the complexity and difficulty of communicating with native creatures, guardians of an indecipherable language. Then, central and mysterious is the role of some individuals – among which the protagonist Avice – who make 'living similitudes', filling with their own body the gaps in communication between colonists and natives, between humans and aliens. There is nothing improper in seeing in this embodiment of meaning a reflection of the shared aspiration to translation that anthropology and science fiction hold. A translation of languages, of worldviews, which requires a joint effort, both interpretative and imaginative, and which is confronted at every step with the limits of the unspeakable – of what cannot be explained or translated. Hence, it is precisely the calling to facilitate encounters, mutual understanding, and exchanges among humans, as well as between humans, non-humans, and other-than-humans. This, together with an awareness of the inherent limits, risks, and illusions of this possibility, is what I believe to be the most stimulating contribution that anthropology and science fiction can offer in envisioning the futures of tourism.

Hence, in the conclusions of this chapter I will draw on the typologies of futures outlined by Bergman *et al.* (2010), particularly focusing on the 'utopia forecast'. By analysing solarpunk as an alternative speculative framework, I will explore how tourism futures can be envisioned through a lens that emphasizes sustainability, inclusivity, and collective ecological well-being.

The MUFANT's Voyages *extraordinaires*

The MUFANT – in English, Museum of Fantasy and Science Fiction is based inside a former school located in a suburban area of Turin. It is a unique museum of its kind in Italy, and in Europe it can only be compared to *La Maison d 'Ailleurs*, the *musée de la science-fiction* (in English, The Science Fiction Museum) founded in 1976 by Pierre Versins in Yverdon-les-Bains, Switzerland. Museums dedicated to science fiction are not numerous worldwide, but the MUFANT signals its peculiarity from its name by presenting itself as MuseumLab. Therefore, it is not only a museum space designed for the conservation and enhancement of science fiction productions, but also a laboratory committed to experimenting, designing, and acting on the spaces outside the museum, within the city.

This paragraph is based on three years of meetings and informal conversations I had between February 2021 and November 2023 with Silvia Casolari and Davide Monopoli, the founders and co-curators of the MUFANT. Our encounters took place in the museum's small thematic library. Surrounded by the masterpieces of the great masters of science fiction – including the series of *voyages extraordinaires* by Jules

Verne – we tried to reconstruct the 'extraordinary' journey of the MUFANT. During an interview I had with them on the 9 December 2021, Silvia told me:

> The museum is managed by a cultural association and since 2002 it has brought together people who are passionate about the genre of fantasy and science fiction, but who also do other things in life and have various skills: psychologists, teachers, collectors. Turin has always had its own tradition on these issues, since the 1970s and we have greatly benefited from this world and the group that had formed around the figure of Riccardo Valla. Today, the museum is open from Thursday to Sunday for the public, the number of visitors is constantly growing, both with regards to Turin citizens and tourists who come from outside. On the other days of the week we work mostly with schools, which come here to visit the museum and participate in various types of workshops. To stay open, a fundamental role is that of internships together with the world of social cooperation. In particular, several people with mental health problems work with us who elsewhere find it difficult to find a job because mental disability is very stigmatised, even more so than physical disability.

In its halls, MUFANT narrates the history and themes of science fiction, spanning from 19th-century origins to contemporary developments. The museum's diverse collection encompasses prints, volumes, and period artifacts from antique collectors, particularly focusing on 19th-century 'proto-fantasy' fiction. A substantial portion comprises cultural products created for mass consumption in the 20th century – toys, comics, costumes, video games, movie posters, miniatures, gadgets, merchandising and packaging – as some sort of materialisation of widespread science fiction in our daily life and recreational experiences. The permanent collection also features original contributions by contemporary artists, manifesting science fiction themes through varied expressive forms. The museum serves as a hub for frequent in-depth studies, hosting conferences, author meetings and study days that delve into the intangible dimensions of this multifaceted material heritage. The MUFANT community, consisting of curators, collectors, enthusiasts, writers and artists, recognizes science fiction as a rich archive of ideas, narratives and objects. This heritage often challenges prevailing public narratives on various social issues, offering cultural criticism and advocating for change. Positioned as a forward-looking heritage, science fiction not only transmits past visions of the future but actively seeks to shape the future itself (Holtorf & Högberg, 2021). Over time, MUFANT has evolved from its original museum role to a hub with a growing interest in designing sociocultural interventions with a significant impact on its surroundings. The museum›s location in a peripheral city area, lacking services and attractions, profoundly influences the community›s activities.

According to Davide Monopoli, this aspect serves as a decisive inspiration, motivating the community to carry out impactful initiatives and engage with the surrounding area. This is a neighbourhood that Davide describes as the 'The Twilight Zone' of Turin, a reference to the title of the famous American science fiction television series created by Rod Serling (1959–1964) focusing on places, stories and experiences that blur the line between real and fantastic (Wolfe, 1997). Using Davide's own words:

> This part of the city is a kind of borderland. I can't say if we arrived here because science fiction has a certain meaning for us, or on the other hand if it acquired it precisely because we are here. In any case, for me science fiction is thinking about possible worlds, which could be here and now. Therefore, it is about seeing what could be there and is not there. It is an existential practice, that is, not loving status quo situations. This is the deep and passionate nature that guides us. We interpret science fiction in this way, with this attitude. Our task as science fiction writers is to tell the story of the other, the alien. Our commitment is manifold, to dissolve prejudices, stereotypes, to deal with the theme that we ourselves have developed and we have become militant in our struggle in favour of the encounter with the other. This has forced us to look more and more at what happens outside the closed spaces of the museum. This is why we have set up what we call the Park of the Fantastic, with the statues of Captain Harlock, Alien, Frankenstein, Wonder Woman and the other characters that can be seen around the museum. This could be defined as a redevelopment project of the area, but we definitely want to avoid things that are just a facade. The aesthetic issue must be a reference for us to the central problem of relating to who lives here, around us. For example, we work a lot with local schools, and then we try to create a connection with families through children. The pop culture and youth aspects of the museum certainly allow us to do this.

The MUFANT's dedicated care for its surrounding area has prompted the City of Turin to consider an energy redevelopment project for the entire block, encompassing the museum and the Park of Fantastic. Proposed interventions include solar panels, 'green walls' on facades, bike-sharing, and electric car charging stations. The objective is to signal environmental concern in a Turin suburb that has endured decades of soil, air, and water pollution.

> We have always thought of ourselves as implementers of small actions of solarpunk tactical urban planning in our neighborhood. I mean, we are certainly all children of cyberpunk, but we are also very careful about what solarpunk can suggest to us, in terms of our ability to imagine a sustainable future, in which new technologies allow us to find a better balance with the environment. We are very close to this way of thinking and do not think that it is in opposition to the cyberpunk approach. Our

vision is not one of opposition between dystopia and utopia; for us, it is complementary, a continuous oscillation.

These words by Silvia refer, in a very interesting way, to the science fiction genre of solarpunk – a movement envisioning a positive, sustainable future with advanced green technology. Hence, this serves as a key to interpreting how the museum relates to the urban environment that surrounds it. This focus has been the primary reason for my collaboration with MUFANT over the last two years. To address the severe repercussions of the COVID-19 pandemic, especially impacting vulnerable populations in urban suburbs, the City of Turin launched the *React to Covid* call in 2022 (Città di Torino, 2022). The museum responded with collaborative activities involving the University of Turin, the Local Health Authority, and city cooperatives. Amid the Anthropocene epoch marking by the irreversible impact of human activities on the environment, the project explored city transformations. Workshops targeted young individuals in vulnerable conditions, including those with mental health challenges and 'Not in Education, Employment, or Training' (NEET) adults. MUFANT conducted guided explorations in Turin's northern suburbs, integrating science fiction narratives to recognize urban space fractures and envision initiatives to mend these tears, guided by the solarpunk perspective.

Faced with the imminent risk of being the last generation on Earth and fuelled by the determination to 'rebel' against extinction, science fiction authors have, for about a decade, embraced a utopian attitude, evident in the emerging solarpunk genre alongside the dystopian perspectives of cyberpunk and post-cyberpunk (Reina-Rozo, 2021). The concept of solarpunk identifies an emerging movement in the field of fantasy and science fiction literature imagining and willing to create sustainable futures characterized by a green energy transition (Williams, 2019). Originating in Brazil and gaining global traction, solarpunk constructs speculative worlds where solar, wind, and tidal energy, technological democratization, and a socially inclusive ecological approach are vital for collective well-being. According to Williams (2019), a peculiar feature of this genre is that the conceptual space of solarpunk preceded its narratives, with much of its recognition rooted in political and social activism rather than literature. This underscores the solarpunk integral role in the struggles and strategies of activists committed to the energy transition. Described as having a 'pedagogical propensity' (Johnson, 2020), solarpunk narratives prompt readers to acquire the knowledge, skills, and creativity necessary for an ecological future. In this perspective, solarpunk writers respond to the call made by the anthropologist and writer Amitav Ghosh in his book *The Great Derangement* (2016). In this work, Ghosh urges a departure from the distant, often apocalyptic visions of climate fiction, encouraging narratives that help us perceive and comprehend the current environmental challenges directly, here, and now.

Epilogue – June 2023, Turin, Italy

Street art at the Bunker, Turin

We're coming to the end of the journey. We are now facing the entrance to what will be the last stage of our urban exploration: the bunker.

Come, follow me... as you will see, among these buildings and industrial warehouses that seem abandoned, there is actually a lot of life. We are in fact inside the old plant of the *Società Italiana di Costruzioni Molle e Affini* (SICMA) – a construction company – which began operating in the 1920s. Beneath the buildings, a large underground bunker was constructed where factory workers sought refuge during the bombings of World War II. That is the reason why the entire area is now called 'the Bunker'.

For many years this entire area was completely abandoned. Then, starting in 2013, the Bunker was gradually transformed into a metropolitan cultural centre. Today, it hosts a contemporary circus school, a theatre company, a cultural circle of African women and several companies focusing on sustainability and recycling themes. One notable initiative is the vast urban garden created by the *Fiësca Verd* association. Covering about 3500 square meters, this group of young people engages in social agriculture activities, viewing gardening as a tool for social integration and inclusion. As you can see, the bunker area hasn't been fully renovated; instead, it embraces greenery and the wild side of nature that reclaimed part of these spaces over the years. This, I believe, makes it the perfect place to read together some excerpts from the *Solarpunk Manifesto*, a text collectively written by authors who identify with this contemporary science fiction genre, expressing their inspiration and vocation here (Solarpunk Italia, 2021):

> *Solarpunk is a literary genre as it is an aesthetics. It is a movement as well: it envisions a better future and constructs operational strategies to attain it. Born in the 2010s, solarpunk represents sentiments and appeals demanding collective, organic, fair, ecological, inclusive progress.*
>
> *Right from its beginnings, it has expressed a complex and open, but clear, political vision: inclusive, feminist, ecologist, utopist, anarchic, organicist. Anticapitalist, antiracist, antipatriarchal, antispecist.*
>
> *Solar. Solar is the prime source and symbol of life, solar is the energy alternative to fossil fuels, it is what already is there, and that we must employ in a sustainable and shared fashion in order to survive. Solar is the light of the day, juxtaposed to rainy, claustrophobic, posturban sceneries of dystopia.*
>
> *Punk. Germs and practice of rebellion. Rejection of the unsustainable, predatory, deadly model of capitalist development, in open contrast to life and vitality, human and not only. A contrarian reaction to dystopic narrative, which no longer provides tools for reacting and declines into conservatism or trendy extinctionism.*
>
> *Solarpunk aesthetics shows not a new, but a renewed sensitivity: environmentalism and anarchism join forces in the search for a future not blacker, but greener.*
>
> *Catastrophe has already happened and now – in artwork of vegetal cities where Art Nouveau meets technology and the art of getting by – we start again together, all of us: persons of every color, origin, physical conditions, age, gender and sexual identity, come together in equal communities. Human and not human animals. Creatures: fauna, flora, Earth.*
>
> *Not compatible with an economy based on consumption and predation, solarpunk preaches no "return to Nature", but pursues instead a conscious progress, whereby science and technology, used in transparent and democratic fashion, will finally allow us to achieve a balance with our biology and our planet.*
>
> *To this an inclusivity just as radical is added, this one indeed born of our times and the momentous rebellion movements of these last years: antiracism and refusal of patriarchy are the foundation for an all-round inclusivity. Towards persons (due to feminist, LGBTQIA*, anti-ableist demands), towards all creatures, towards the world, with the refusal to split ontologically the human beings from their ecosystem.*
>
> *We believe that solarpunk is an ongoing process, not a defined phenomenon.*
>
> *Our keyword, as we write solarpunk and we write about solarpunk, is hope.*

Conclusions: Envisioning Tourism Futures Through a Solarpunk Utopian Scenario

Approaching conclusions, I would now like to explain what, in my opinion, are the most direct implications for the futures of tourism resulting from the intertwining of anthropology and science fiction that I have outlined so far.

To this end, it is important to revisit the meaning of the expression 'urban exploration' that I have used since the beginning of this contribution and that I have also mentioned in its title. 'Urban exploration' encompasses a broad spectrum of experiences related to discovering and visiting abandoned, ruined, or otherwise little-known places by city inhabitants or those excluded from them. This social practice, carried out individually or collectively, can take on various meanings – political (e.g. challenging policies of privatisation and segmentation of urban space), aesthetic-artistic (e.g. creating photographic reportages) or touristic. In the touristic context, Robinson (2015) highlighted how urban explorations can be seen as a fusion of ecotourism and critical heritage tourism. It is a practice that offers an unfiltered view of the past, emphasizing the social conflicts, failures, and contradictions of industrial and urban development. In this sense, urban exploration is an experience that goes 'beyond tourism', challenging the predatory consumption of the territory's resources and the illusory construction of official narratives of history.

I believe that urban exploration provides the most suitable framework for understanding the potential that science fiction holds for the futures of tourism in the Anthropocene, or rather, the Haraway's (2016) *Chthulucene*. Haraway coined the term Chthulucene, challenging the anthropocentrism inherent in the concept of the Anthropocene. Haraway's conceptualization also contributes, at the terminological level, to the 'ontological turn' inspired by the works of Philippe Descola (2013), Bruno Latour (2017), and Eduardo Viveiros de Castro (2015). This intellectual shift encourages to deconstruct the nature-culture dichotomy and prompts a re-evaluation of the fundamental idea of 'environment' (Tassan, 2021). For Haraway, it is crucial to escape the salvific and apocalyptic projections of the future, and immerse oneself deeply in the present, in a configuration of chthonic time-space, forming unpredictable kinships with other living creatures. In her effervescent linguistic creativity – a characteristic likening her theorisations to science fiction- Haraway invites us to 'wander' in the relationships with non-human 'companion species'. The connections between solarpunk science fiction and Donna Haraway's work are more than evident. The essay on the Chthulucene ends not by chance with an experiment in speculative narration written by Haraway, the story 'Children of the compost' (Haraway, 2016). On the other hand, there are numerous solarpunk narratives that feed on Haraway's categories of thought and lexicon. It is enough to mention, for example, the story

Children of Asphalt by Phoebe Wagner (2021), which uses the concept of kinship between humans and other-than-humans bringing it into the context of multispecies cities of the near future.

Framing *solarpunk tourism* as a utopian scenario, I draw on Bergman *et al.* (2010) and their typology of futures studies, positioning solarpunk within the utopia type of forecast. Rather than a fixed vision of the future, the solarpunk approach embraces the anticipatory capacity of futures studies to depict a world where tourism functions within an environmentally conscious, equitable, and technologically integrated framework. *Solarpunk tourism*, then, can be seen as a speculative model for sustainable urban travel, where speculative fiction converges with anthropology to explore alternative modes of engagement with the city. It aligns with the tradition of utopian thought in futures studies, not as an unattainable ideal but as a guiding direction for possible transformations within tourism in the Anthropocene era. Hence, here is my vision for the futures of tourism: a *solarpunk tourism*, capable of 'wandering' in multi-species urban environments, exploring the twilight zone where the real and the fantastic merge, and where relationships between human and other-than-human creatures develop. It is an experience 'beyond tourism' where anthropologists and science fiction scholars collaborate to explore poetically, politically, and aesthetically what Ghosh (2016: 33) called the 'disturbing intimacy of our relationship with the non-human'.

This kind of tourism futures are a concrete and unavoidable possibility in our post-pandemic era. As Sennett (2018) wrote, one of the main ethical and political challenges of the present is to 'open' the city by imagining and contributing to creating 'open cities'. Open cities allow a social life different from the predominant model of the 'closed city', which is strongly irregular, segmented, and subjected to many control devices. The prolonged periods of lockdown and the 'social distancing' due to the need to fight the COVID-19 pandemic have further emphasized the interconnectedness of ecological balance, economic and social inequalities, and the coexistence between human and non-human life forms. Meschiari (2020) rightly argued that doing research and writing in the Anthropocene also means acquiring a new language, new symbols, and different ways of telling stories. The same, I believe, applies to tourism, for scholars, practitioners, and all those interested in exploring its futures in the Chthulucene.

Acknowledgements

I sincerely thank Davide Monopoli and Silvia Casolari, founders, and co-curators of MUFANT, Paolo Bertetti, who is its scientific coordinator, and all the museum staff for having welcomed me and always supported and facilitated my research work.

References

Adams, D. (1979) *The Hitchhiker's Guide to the Galaxy*. Pan Books.

Appadurai, A. (1996) *Modernity At Large: Cultural Dimensions of Globalization*. University of Minnesota Press.

Bauman, Z. (1998) *Globalization: The Human Consequences*. Columbia University Press.

Bergman, A., Karlsson J.C. and Axelsson, J. (2010) Truths claims and explanatory claims – an ontological typology of futures studies. *Futures* 42 (8), 857–865.

Clarke, A.C. (1968) *2001: A Space Odyssey*. Hutchinson.

Collins, G. (2021) *All Tomorrow's Cultures: Anthropological Engagements with the Future*. Berghahn Books.

Dei, F., Dimplflmeier, F. and Vietti, F. (eds) (2023) Le meraviglie del possibile. Antropologia e fantascienza, Rivista di Antropologia Contemporanea (RAC), 1/2023: 5–15.

Descola, P. (2013) *Beyond Nature and Culture*. University of Chicago Press.

Ghosh, A. (2016) *The Great Derangement: Climate Change and the Unthinkable*. University of Chicago Press.

Hao, J.F. (2015) Folding Beijing (北京折叠). *Uncanny Magazine*. Available online at https://www.uncannymagazine.com/article/folding-beijing-2/

Haraway, D. (2016) *Staying With the Trouble: Making Kin in the Chthulucene*. Duke University Press.

Holtorf, C. and Högberg, A. (eds) (2020) *Cultural Heritage and the Future*. Routledge.

Johnson, I. (2020) "Solarpunk" & the pedagogical value of Utopia. *Journal of Sustainability Education* 23, Ecomedia Literacy.

Latour, B. (2017) *Facing Gaia: Eight Lectures on the New Climatic Regime*. Polity Press.

Leed, E.J. (1991) *The Mind of the Traveler*. Basic Books.

Mathews, G. (2011) *Ghetto at the Center of the World: Chungking Mansions, Hong Kong*. University of Chicago Press.

McCarthy, H. and Ashmore, D.J. (2021) *Leiji Matsumoto: Essays on the Manga and Anime Legend*. McFarland.

Meschiari, M. (2020) *Anthropocene Fantastico*. Armillaria.

Miéville, C. (2009) *The City & the City*. Random House.

Miéville, C. (2011) *Embassytown*. Pan Macmillan.

Miller, S.J. (2018) *Blackfish City*. Ecco Press.

Okorafor, N. (2014) *Lagoon*. Hodder & Stoughton.

Park, R.E. and Burgess, E.W. (eds) (1925) *The City*. University of Chicago Press.

Reina-Rozo, J. (2021) Art, energy, and technology: The Solarpunk movement. *Social Justice and Peace* 8 (1), 47–60.

Remotti, F. (1990) *Noi Primitive. Lo Specchio Dell'antropologia*. Bollati Borighieri.

Robinson, P. (2015) Conceptualizing urban exploration as beyond tourism and as anti-tourism. *Advances in Hospitality and Tourism Research* 3 (2), 141–164.

Salazar, N. and Glick Schiller, N. (eds) (2014) *Regimes of Mobility: Imaginaries and Relationalities of Power*. Routledge.

Sennett, R. (2018) *Building and Dwelling: Ethics for the City*. Penguin Books.

Shamir, R. (2005) Without borders? Notes on globalization as a mobility regime. *Sociological Theory* 23 (2), 197–217.

Smith, V.L. (1978) *Hosts and Guests. The Anthropology of Tourism*. Basil Blackwell.

Solarpunk Italia (2021) *Solarpunk Manifesto*. Available online at https://solarpunk.it/what-is-solarpunk-a-manifesto/.

Stover, L.E. (1973) Anthropology and science fiction. *Current Anthropology* 14 (4), 471–474.

Tassan, M. (2021) Rethinking environmental justice. Anthropological perspectives on environments, nature and inequalities in the Anthropocene era. *Antropologia* 8, 2 NS, 11–36.

Urry, J. (2000) *Sociology Beyond Societies: Mobilities for the Twenty-first Century.* Routledge.

Vietti, F. (2023) Salvare il futuro. Antropologia, fantascienza e altri mondi possibili. In C.S. Guerzoni and V.L. Toro Matuk (eds) *Degni di umanità. Verso un'estetica antropologica* (pp. 145–170). Mimesis.

Viveiros de Castro, E. (2015) *The Relative Native: Essays on Indigenous Conceptual Worlds.* Hau Books.

Wagner, P. (2021) Children of asphalt. In C. Rupprecht, D. Cleland, N. Tamura, R. Chaudhuri and S. Ulibarri (eds) *Multispecies Cities: Solarpunk Urban Futures* (pp. 140–145). World Weaver Press.

Wells, H.G. (1895) *The Time Machine.* William Heinemann.

Wells, H.G. (1898) *The War of the Worlds.* William Heinemann.

Wheeler, T. (1973) *Across Asia on the Cheap.* Lonely Planet.

Williams, R. (2019) 'This shining confluence of magic and technology': Solarpunk, energy imaginaries, and the infrastructures of solarity. *Open Library of Humanities* 5 (1), 1–35.

Wolfe, P. (1997) *In the Zone: The Twilight World of Rod Serling.* Bowling Green State University Popular Press.

Part 3

Views and Insights from Independent Researchers and Practitioners

11 Tourism, Urban Population and *Othering*: Critical Reflections and Forecasts for Italian Cultural Cities

Sarah Gainsforth

Introduction: The Italian Urban Space

Cities used to be surrounded by walls. Today, where walls were built, cities have incorporated them into the expanding urban fabric. The rise of urban populations and the increasing movement of temporary populations through cities have radically altered the social fabric of urban spaces. The contemporary metropolitan urban form is indeed quite different from that of a traditional settlement, which coincided with the resident population. Cities attract a variety of different populations; they are traversed by mobile and temporary populations that continually reshape the relation with the land. Quite prophetically, in 1993, the leading Italian sociologist Guido Martinotti wrote:

> Of the total population that gravitates to the city, the proportion of the population that uses it, but is not "responsible" for it, tends to increase and perhaps will eventually outnumber even the total population that uses the city but also inhabits it, and is therefore responsible for it. (Martinotti, 1993: 166)

In his seminal book *Metropoli, La nuova morfologia sociale della città* (*Metropolis, The New Social Morphology of the City*), Martinotti (1993) describes the different social formations connected in contemporary urban space, layers adding to and transforming the traditional town into a new entity: the modern urban metropolitan area, a functional system more than a spatially defined structure. The following quotes are from

Martinotti's paper *The New Social Morphology of Cities* synthesising a lecture held in Vienna in 1994. To explain 'the great mutation' affecting contemporary cities, Martinotti (1994: 4) proposed to 'represent schematically various types of urban morphologies by using a simple combination of *four populations* differentiating out in successive phases'. The *inhabitants*, or the population *living* in the city, which in the traditional city largely coincided with the population *working* in the city; the *commuter population*, whose growth took place in the USA from the Twenties and in Europe after World War II. The commuter population is concentrated in the city centre and remains mostly separated from the rest of the city population, Martinotti argues, and has led to the formation of a new entity: the *first-generation metropolis,* largely based on *commuting basins* and embodied in the concept of *metropolitan area*. Further differentiation of populations arose from innovation within the transportation system: 'the increased mobility of people, combined with the availability of greater income and leisure, allowed the differentiation of a *third population* in our diagram, the *city users*' (1994: 5).

Martinotti notes that 'this is a swelling population that has increasing effects on the structure of cities and actually uses them in a rather uncontrolled way' (1994: 5). The proportion between inhabitants, commuters and city users varies among cities – Venice, already at the time of Martinotti's writing, being an extreme case of disproportion in favour of city users compared to other populations. Martinotti identified a fourth metropolitan population made up of *metropolitan businessmen*, a relatively small but growing group according to him, a product of the service industry (similarly to city users) disposing of a 'considerable availability of both private and corporate money' (1994: 6), carrying out activities that require intense and repeated face-to-face interaction. This *transnational middle-class* population living *between cities* signals the globalisation of metropolitan centres. One of the major effects of globalisation and of the growth of this fourth population, Martinotti argues, is the increasing homogenisation of cities that once 'embodied the specificity of local culture and traditions' (1994: 7). According to Martinotti, the presence of these four populations, characterises the third-generation metropolis, which is still emerging.

Aside from the description of the four populations and of the new types of urban morphology they give rise to, a second aspect of Martinotti's theory, in fact a cornerstone in *Metropoli*, is worth recalling because of its strong relevance to the current state of cities in Italy: the inability to assess urban transformations by changing populations due to the inadequacy of traditional statistical data based mainly on residency. The size of the population of city users is growing, Martinotti writes, but it is difficult to assess. 'Precisely because all our collective cognitive apparatus is geared to a traditional city that is undergoing a profound mutation, and statistics still deal mainly with *inhabitants,* to a smaller degree with *commuters,* but practically in no way with *users*' (Martinotti, 1994: 5).

In 2016, the Italian National Institute of Statistics (Istat) used an experimental method to calculate the incidence of worker and student commuters (thus excluding tourists) in Italy's major cities during the day, for the first time: on average, their number is 20% of the resident population. Although the data may be overestimated, the number of commuters in Rome is equivalent to one-third of the resident population; in Milan it is more than half.

The inability to clearly assess changes and shifts in balance of different populations that impact the urban morphology bears serious consequences. Indeed, *Metropoli* begins with the tale of the incorrect use of the 1981 Italian statistical census data, apparently recording a decrease in the population of the largest Italian cities. The incorrect reading of this data, driven by Istat's grouping of municipalities according to their demographic size, fed interpretations of counter-urbanization, de-urbanization, suburbanization, outright 'death of the city', and the argument that people were 'returning to live in the countryside'. However, when introducing the fundamental distinction between metropolitan and non-metropolitan municipalities, the data revealed not a 'death of the cities', but what Martinotti framed as 'the key dynamic of metropolitan transformation' (Martinotti, 1993: 48): a faster growth of metropolitan municipalities below 100,000 inhabitants, compared to non-metropolitan municipalities of the same demographic size. In brief, the key dynamic of contemporary metropolitan transformation is not a return to the countryside, a rhetoric which gained popularity once again in 2020 during the pandemic, but the growth of large metropolitan areas and the decline of the population in the city centres.

In the past decades, the redistribution of residents has downsized the demographic weight of the urban cores in cities in Italy. In Rome, for instance, the population has not varied significantly from the early 1970s to the early 2000s; however, the historic centre and the semi central historic districts have lost about 800,000 residents, almost one-third of the population. Hence, the number of residents in the city centre has dropped from 64% of the total municipal population in 1971 to 35% in 2016. Meanwhile, the population of Rome's hinterland doubled and is now close to 1.5 million (Crisci, 2018). So, the 'depopulation' of Italian city centres is nothing new. However, this phenomenon has intensified in the past decade due to a notable increase in tourism.

Looking toward the future, these trends will likely accelerate. According to Bergman *et al.* (2010), understanding urban futures requires acknowledging the ontological nature of population projections and its interplay with tourism. Urban studies must account for these evolving dynamics, especially in cultural cities that are also tourism destinations where long-term residency is being outpaced by the influx of temporary urban populations, primarily tourists, commuters, and transient workers. Urban planning will need to address the competing interests of residents, tourists, and temporary inhabitants, all while mitigating the risk of exacerbating inequalities and environmental degradation. Inayatullah's (1998) critical futures research provides a framework emphasizing the need to consider the deeper causal forces shaping urban spaces. The role of tourism and mobility is not simply a matter of economic growth but also one of cultural identity and spatial equity. If the current trajectory continues, the lines between permanent residents and temporary users will continue to blur, creating tensions over resource allocation, public space usage and city governance.

This chapter explores how tourism is reshaping the Italian urban space and discusses future scenarios of tourism in urban spaces, including both utopian and dystopian forecasts. Urban planning must rethink its assumptions about population and spatial justice, considering the diverse needs of both permanent and transient residents.

The Role of Tourism

Over the past three decades, tourism has reshaped Italian cultural cities at an unprecedented new speed, a 'dramatic dimension' added to the constant processes of 'creation, destruction and transformation of the inhabitable land', Martinotti (1994: 2) writes. Still in the Nineties, Martinotti understood the growing relevance of tourism in the urban landscape; he advocated for the introduction of a degree course in Italian universities in tourism sciences, which he coordinated at Milano-Bicocca. The speed at which we are connected to different points 'in the matrix of persons, social units, and events' (Martinotti, 1994: 2) witnessed a leap in scale with new routes and means of low-cost transportation. The number of travellers worldwide has soared: international travellers increased from 25 to 674 million from 1950 to 2000; then, in just 15 years, they doubled to 1.3 billion in 2017 and reached 1.5 billion in 2019 (UNWTO, 2020). The rise of a new hospitality market created with injections of billions of dollars of Silicon Valley venture capital into digital platforms such as Airbnb has led to another leap in scale in tourism over the past decade, with the growth of short-term tourist home rentals.

In 2019 homes on Airbnb exceeded those rented to residents in the centres of six Italian cities: Bologna, Florence, Naples, Palermo, Rome and Venice, according to data from Inside Airbnb analysed by the geographer Filippo Celata (Celata & Romano, 2020). Neighbourhood shops have disappeared, prices have risen and the continuous flow of tourists has congested streets and squares, giving rise to issues of liveability strongly felt by the remaining residents. In fact, increased mobility is often not matched by improvements in physical infrastructure, services and public spaces.

In many Italian cities tourism is no longer merely an important sector of the urban economy, but it is its central driving force – according to Istat 'the enhancement of the cultural heritage of cities is a key determinant of the urban economy, if not, in many cases, the only relevant one' (Istat, 2020: 9). This shift is in part the result of the economic recession that hit Italy after the 2008 crisis, when Italy's industrial output shrank by 25% (Istat, 2014) and aggregate domestic demand crumbled under fiscal policy measures which entailed cuts in state funding to cities, totalling 12 billion euros from 2011 to 2018 (Anci, 2018). In this scenario tourist expenditure, 'Italy's oil', was widely viewed as a means for economic compensation.

Tourism has thus become an important sector of the urban economy and has been actively promoted by municipalities as an urban development strategy. The result has been further impoverishment: jobs in the touristic sectors have been created, often substituting qualified higher wage labour. Between 2004 and 2011 almost half of the employment growth in the tertiary sector in Italy was due to the development of hotel, restaurant and domestic services (Banca d'Italia, 2021): low-wage, precarious labour, which grew as more qualified sectors shrank. In Rome, a city traditionally lacking an industrial sector, the sectors that recorded the most significant increases between 2008 and 2019, both in absolute and relative terms, were personal and collective services and business services, closely followed by hotels and restaurants (Comune di Roma, 2019). Indeed, the Bank of Italy notes that over a period of nearly two decades, the performance of the value added per capita in the Capital has been worse than that of other major cities Italian and European cities, a consequence of an unsatisfactory dynamic of product per person employed. The public sector, large enterprises and investments have shrunk, while there has been a rapid growth of employment in low knowledge-intensive services, partly due to the sharp increase in tourist flows. This has resulted in a significant reduction of specialization in knowledge-intensive services and a sharp expansion of lower-skilled occupations (Banca d'Italia, 2023: 1).

Despite the evidence of the rising social, environmental and economic costs of over-tourism, administrations in cities that have already reached their capacity continue still today to promote tourism as a means of economic growth. In Milan, beginning with the bid to host the 2015 World Expo, considered by some to be the main driver of the city's 'renaissance', the promotion of tourism has become a key component of the administration's urban growth strategy. But with tourism inequality has grown disproportionately. Between 2015 and 2021 incomes and wages in the Milan metropolitan area have grown by 12 and 13% respectively, while house prices have risen by 41% and rents by 22%, according to the Milan Metropolitan Affordable Housing Observatory (OCA), which titled its annual report 'Not a City for Working People' (OCA, 2023).

The rent gap generated by temporary stays has changed the property market in the main Italian city centres, where there is no longer a supply of ordinary long term housing. As of 2022, the issue of housing affordability has gained prominence in the national debate, as prices and rental demand rose after 2020. The demand for rental housing has led to a rapid increase in price, also boosted by short tourist rentals. Prices at the high end of the real estate market have spurred an increase also in prices in the low and mid-range markets; hence the range of values in the extreme areas of the city, the centre and the suburbs, has narrowed. In short, the private urban real estate markets (the real estate market in Italy is predominantly private) address an increasingly select target of national and international *temporary* tenants, leading to an increase in specific demand for housing and services related to a temporary permanence in the city. But the increase in mobility has reduced the ability to inhabit cities for all those categories of people who are not at the top of the wealth ladder. In cities such as Florence, long-term rentals are disappearing altogether. And while tourism is recovering after the pandemic, strongly driven by visits to cities of art which in 2022 grew 275% compared to the previous year (Banca d'Italia, 2023a), something else is happening as well: we are witnessing the rise of medium-stay rentals intermediated by new, international, expanding digital platforms, such as DoveVivo and Jovy, in the role of intermediaries that manage thousands of apartments.

Forms of temporary living, other than traditional migration and tourism, appear to be fast-growing. If public administrations ignore emerging trends, private real-estate operators have already picked up on them. Or are they inducing them? How much are new mobility patterns, presented as lifestyle trends, induced by the housing market and by an increasingly precarious job market? In other words, how much are new mobility patterns determined by the housing and the job markets, and how much are they actually 'lifestyle' choices? Today there is little data to understand and substantiate the alleged increase of demand for medium-stay rentals, and what is described as a lifestyle turn towards increased mobility. However, 'multilocality, that is multilocal living characterised by the interplay between mobility and stability, does appear to be a social phenomenon on the rise' (Rainer *et al.*, 2021), for example in regions such as the Dolomite area in the Italian Alps that have witnessed a decline in forms of traditional tourism, also thanks to the availability of second homes' (Elmi & Perlik, 2014).

Of course, with inadequate public data sources to understand emerging mobility patterns, it is difficult to know exactly who is driving the change, once again. In cities, judging by the reshaping of the housing market and of the urban morphology by short-term rentals catering to tourists, what is more likely the case is a repositioning of the *supply* side to attract temporary populations of non-residents in Italian cities.

The arrival of new inhabitants, albeit temporary, does have positive effects in social and cultural terms. The problem lies in other, negative, outcomes such as gentrification, sociospatial polarization and increased inequality that a richer, more mobile population can produce. A similar outcome could be impacting non-urban areas where lifestyle migration and multilocal living arrangements, also induced by climate change and remote working, could be determining selective movements and settlement of population, based on socioeconomic status, relational and symbolic resources. Indeed, both in cities and in non-urban areas the effects of gentrification are often sought by municipalities, who leverage attractiveness to foster urban development, using gentrification as an urban growth strategy, ignoring its implications in terms of costs. But temporary populations, made up of different categories whose boundaries are increasingly blurred also as a result of the reshaping of the real estate market, contribute to the expulsion of residents – who, by the way, pay for public services to function. In a worrying scenario of demographic decline and ageing resident population, Italy desperately needs new residents – not just temporary ones. The future of tourism and of the contribution it can provide depends greatly on urban policies: cities, as well as non-urban areas, must cease to be mere 'consumption devices' and return to be places to inhabit.

Conclusions: The Future of Tourism in Italian Cultural Cities Through Dystopian and Utopian Forecasts

After the COVID-19 pandemic, tourism in Italy has risen again. But it is also evolving and changing shape. Traditional categories of tourism (leisure, cultural, business, religious and so on) are still valid, but something else is also taking place in the space between the most permanent and the most temporary categories of city dwellers, between the 'resident' and the 'tourist' populations. As Martinotti put it, 'as it happens in all periods of deep structural mutation, the old and the new are highly mixed, in reality as well as in the minds of men, and it is difficult to severe one from the other' (Martinotti, 1994: 3). Within the spectrum of permanent and ultra-temporary residents and users, boundaries between different populations seem to be increasingly blurred by new selective mobility patterns. These need further investigation in order to understand what is driving mobility and temporary living trends, how and why the conditions of place 'livability' are changing, who are they catering to, and what spatial and social impacts we should expect.

Art cities and popular destinations are witnessing the growing negative impacts of tourism, such as congestion, shortages of affordable housing, loss of cultural identity and environmental degradation. The so-called phenomenon of overtourism has emerged as a significant threat the sustainability of cities. At a time when it is imperative to foster a new sustainable economic model within ecological planetary boundaries for the so-called green transition, it would be interesting to have more data on the ecological footprint of tourism in urban cultural destinations such as Rome. The ecological footprint measures the amount of productive land necessary to sustain a community's lifestyle (production, consumption, waste). If this measure exceeds the actual surface area, which is 'weighed' according to its biological capacity, the community has a deficit. To make up this deficit, tourist destinations subtract resources from other locations, and from the system in general, and input waste. The problem of sustainability is thus closely connected with that of equity. The ecological footprint in relation to tourism has been explored in leisure destinations such as small islands, and the findings are not comforting. According to the Strategic Plan for the Sustainable Development of the Pelagic Islands, a study conducted by the Department of Urban Planning at IUAV University of Venice, the land area required to produce the resources consumed each year in Lampedusa, a small island off Sicily, is 4.5 times the island because of the influx of tourists.

In this light, the future of Italian cultural cities can be envisioned through both utopian and dystopian forecasts, drawing on Bergman *et al.* (2010)'s typologies of futures studies and Inayatullah's (1998) critical futures research approach for alternative futures.

A dystopian tourism future for Italian cultural cities

A dystopian future, already partially unfolding, sees tourism's unchecked expansion leading to a 'survival of the richest' model, where commodification of public spaces, housing, and heritage accelerates displacement and social stratification. Policies such as entry fees and access restrictions, as introduced in Venice, risk deepening inequalities rather than addressing the root causes of resource depletion. Instead of fostering a more sustainable balance, such measures may reinforce tourism-driven exclusivity, where access to cultural and natural assets becomes contingent on purchasing power. This trajectory and forecast caution that if mobility governance prioritises economic imperatives over social equity, cities risk exacerbating spatial injustices and reinforcing exclusionary urban dynamics.

A utopian tourism future for Italian cultural cities

Conversely, a utopian future is conceivable – one where tourism is regulated through policies that prioritise urban livability, social justice, and ecological balance. Achieving this requires a paradigm shift in governance, where decision-making is guided by equitable access to public space, affordable housing, and sustainable mobility. As Martinotti (1994) observed, existing administrative and analytical tools often struggle to capture evolving urban dynamics. Reclaiming control over public data – rather than relying on private companies monetizing mobility information – would empower policymakers to develop evidence-based strategies for managing tourism's impact. Furthermore, a 'just transition' approach would ensure that tourism does not monopolise shared resources, but instead contributes to local well-being through redistributive mechanisms and inclusive urban planning.

Ultimately, the balance between tourism, residents, and mobility in Italian cultural cities hinges on the capacity of public institutions to regulate access and consumption in a way that does not deepen existing inequalities. Without proactive intervention, the dystopian future – marked by commodification, social exclusion, and environmental depletion – will continue to unfold. However, with a governance model that integrates sustainability, justice, and democratic data management, a more equitable and resilient urban future remains within reach.

References

Anci (2018) Audizione Anci, Commissioni Bilancio Riunite del Senato della Repubblica e della Camera dei Deputati, Legge di Bilancio 2018. Available at https://www.fondazioneifel.it/ifelinforma-news/item/9067-audizione-anci-sulla-legge-di-bilancio-2018 (accessed 26 April 2024).

Banca d'Italia (2021) Aimone Gigio, L. and Camussi S.A.M., *Cambiamenti nella struttura qualitativa dell'occupazione*, Questioni di Economia e Finanza (Occasional Papers), Banca d'Italia, n. 603, 2021. Available at https://www.bancaditalia.it/pubblicazioni/qef/2021-0603/index.html?dotcache=refresh (accessed 26 April 2024)

Banca d'Italia (2023) Bronzini, R. Bolis, M. Daniele, F. Di Carmine, C. Leva, L. Montaruli, F. Romito, E. Ruggieri, D. and Scarinzi E. *L'economia di Roma negli anni duemila. Cambiamenti strutturali, mercato del lavoro, diseguaglianze.* Questioni di Economia e Finanza (Occasional Papers) n. 793, 2023. Available at https://www.bancaditalia.it/pubblicazioni/qef/2023-0793/index.html (accessed 26 April 2024)

Banca d'Italia (2023a) Indagine sul turismo internazionale – 2022. Statistiche. Available at https://www.bancaditalia.it/pubblicazioni/indagine-turismo-internazionale/index.html (accessed 26 April 2024).

Bergman, A., Karlsson, J.C. and Axelsson, J. (2010) Truths claims and explanatory claims – an ontological typology of futures studies. *Futures* 42 (8), 857–865.

Celata, F. and Romano, A. (2020) Overtourism and online short-term rental platforms in Italian cities. *Journal of Sustainable Tourism* 30 (3), 1–20. https://doi.org/10.1080/09669582.2020.1788568

Comune di Roma (2019) *Annuario Statistico, Cap 2 Le caratteristiche dell'occupazione.* Available at https://www.comune.roma.it/web-resources/cms/documents/Annuario_statistico_2019.pdf (accessed 26 April 2024).

Crisci, M. (2018) Fine dello sprawl a Roma? La capitale verso una nuova fase di sviluppo urbano. In A. Coppola and G. Punziono (eds) *Roma in Transizione. Governo, Strategie, Metabolismi e Quadri di vita di una Metropoli* (pp. 66–78). Planum Publisher.

Elmi, M. and Perlik, M. (2014) Dal turismo alla residenza multilocale? *Journal of Alpine Research*, 102–3 2014. https://doi.org/10.4000/rga.2600

Inayatullah, S. (1998) Causal layered analysis: Poststructuralism as method. *Futures* 30 (8), 815–829.

Istat (2014) Evoluzione Ciclica della Produzione Industriale e del Fatturato in Italia e in Europa. Available at: https://www.istat.it/it/files/2014/02/capitolo-1.pdf (accessed 26 April 2024).

Istat (2020) Turismo D'arte in Area Urbana, una Proposta di Indicatori. Available at https://www.istat.it/it/archivio/252531 (accessed 26 April 2024).

Martinotti, G. (1993) *Metropoli, La Nuova Morfologia Sociale Della Città*. Bologna, Il Mulino.

Martinotti, G. (1994) The New Social Morphology of Cities, UNESCO/MOST Wien, 10–12 February 1994. Available at http://www.laboratoriorapu.it/lpu_sito/download/Materiali_Papers/07d_08.04.16_New%20social%20morphology-Martinotti.pdf (accessed 26 April 2024).

Osservatorio Casa Abbordabile di Milano Metropolitana (OCA), Dipartimento di architettura e studi urbani, Politecnico di Milano (2023) *Non è una città per chi lavora. Costi abitativi, redditi e retribuzioni a Milano. Primo rapport di ricerca* OCA *sull'abbordabilità della casa*. Available at https://oca.milano.it/wp-content/uploads/2023/11/OCA_Report_2023.pdf (accessed 26 April 2024).

Rainer D. *et al.* (2021) Multilocal living and spatial development, Positionspapier aus der ARL, n. 123, Verlag der ARL - Akademie für Raumentwicklung in der Leibniz-Gemeinschaft, Hannover. Available at https://nbn-resolving.de/urn:nbn:de:0156-01231 (accessed 26 April 2024).

UNWTO (2020) International tourism highlights – 2019 edition. Available at https://www.e-unwto.org/doi/pdf/10.18111/9789284421152 (accessed 26 April 2024).

12 The Space of Young Residents with a Migrant Background as Cross-Border Guides: The Project 'Rural Migrantour: Paths of Recovery'

Jana Milovanovič

Context and Project Rationale

International migrations are increasingly taking shape as phenomena characterized by transnational aspects that create strong links between different territories and cultures. From these considerations, the idea of 'Migrantour' was developed in 2010 in Turin, Italy, by Viaggi Solidali cooperative, a responsible tourism tour operator that had first matured this proposal. The project aims to promote a form of responsible zero-mile tourism that has, as its protagonists, fellow citizens who have experienced migration in their biographies or family histories. Migrantour network now operates across Europe, offering travellers and residents a new perspective on multicultural environments. While initially concentrated in urban settings, extensive research conducted by individual Migrantour entities has unveiled the richness and complexity of rural areas as fertile grounds for cultural convergence. These rural expanses not only harbor untold stories and overlooked histories but also serve as crucibles for the mingling and blending of diverse cultures.

The strategic partnership project *Rural Migrantour – Paths of Recovery*,[1] implemented under Erasmus Plus Strategic partnership programme between March 2021 and September 2023, enabled the expansion of the previous Migrantour project as an established European network of intercultural walks to the rural regions. This expansion aimed to facilitate recovery during/after the COVID-19 pandemic, particularly addressing its impact on rural youth in vulnerable contexts facing geographical, social and cultural obstacles. While the original Migrantour initiative centred on European urban spaces, Rural Migrantour ventures into rural territories, where it seeks to document and disseminate narratives related to minority cultures, border dynamics and contemporary migration issues. Drawing on testimonies from local inhabitants, the project examined the sociocultural impacts of borders, the experiences of marginalized communities, and the complexities of rural migration landscapes.

The project sought to empower young individuals by creating opportunities in the field of responsible tourism, fostering upskilling and a sense of belonging. Social inclusion and active citizenship were developed through the creative use of cultural heritage, considering peripheral, depopulated – or otherwise affected – rural regions as vibrant creative spaces for cultural dialogue and diversity. These territories were recognized for their significant potential to engage locals as intercultural guides, activating their capabilities in the field of responsible tourism. The project involved five partner organizations from Slovenia, Greece, Italy, the Netherlands and Bulgaria.

Borders are where the wars start, as Primo Levi, an Italian Jewish writer and Holocaust survivor, wrote in his work The Wrench (orig. La chiave a stella). However, they are also bridges – sites of ongoing cultural exchange. Peripheries produce complicated geographies, often marked by limbo, forms of exception, ambiguous political realities and continuous transformation (Dikomitis, 2005; Lundén, 2006; Navaro-Yashin, 2012). In this regard, the project aimed to reshape the concept of European cultural heritage by focusing on the voices and stories of the geographical and cultural margins, specifically the peripheries and borderlands, whether open or regulated by the Schengen regime, contemporary or historical. It primarily supported communities that have experienced the consequences of historical events, policy changes and border restrictions. The emphasis on communities at geographical and cultural peripheries allowed their voices to be heard and their experiences were documented. In doing this, the project contributed to a more comprehensive understanding of European history and heritage and to the construction of a novel future scenario for next generations. Additionally, the project aimed to transmit oral histories to younger generations, ensuring the continuity of knowledge and the preservation of cross-border solidarity and cooperation, beyond any challenging cultural, historical, or political process of othering.

The *Rural Migrantour* project has selected five peripheral and/or borderland microsites: Gorjanci-Žumberak (Slovenia-Croatia), Svilengrad (Bulgaria), Kythera (Greece), Camini (Calabria, Italy) and Paraloup (Piemont, Italy), to investigate the impact of policies on the people living in these marginal regions, past and present. The continuous shaping and reshaping of these selected areas has often been accompanied by war violence, the deportation of populations, desertion, migration and illegal border crossings. All these factors have inevitably resulted in many traumas and disrupted the traditional patterns of migration, agriculture and commerce. The project's focus on engaging oral history testimonies and sharing the narratives of individuals affected by borders/peripheries had the potential, as we hoped, to contribute to personal and societal healing. To transmit these stories to future generations in an engaging and attractive way, the project employed creative multimedia storytelling tools and involved young individuals in the co-creative production processes. As such, the project outcomes help to envision an inclusive cross-border cultural tourism in alignment with Bergman *et al.*'s (2010) typologies of futures, more specifically as a utopian forecast.

The Five Project Locations

Camini, Calabria, Italy

Camini is a small village in Calabria, Italy. For decades, Locride area – and Camini – was hollowed out by emigration, a direct result of the extreme poverty and lack of job opportunities. Over this same period, the area was also blighted by the activities of organized crime groups, which had a negative impact both on the wider region and beyond. Despite the beauty of the place, tourism struggled to take hold, leading to a significant exodus. Nowadays, thanks to the activities of civil society organizations and growing community resilience, the situation is changing. In 1999, a social cooperative project *EuroCoop – Jungi Mundu*[2] was established in order to favor job placement for weak members of society, evolving in 2011 to offer a fully validated reception service for migrants, aiming mainly to guarantee independence and integration of asylum seekers and refugees. The project paved the way for the current operations center 'Jungi Mundu' – which means 'Join the World' in the local dialect. It facilitates asylum seekers' reception and integration, adopting a scattered hospitality approach in contrast to the transitory way of life that typifies the crowded migration centres in many larger cities. While the overall pattern remains uneven across the region, the village of Camini itself is undergoing something of a renaissance. It has experienced social and economic growth as a result of its reception and integration of third-country nationals programme, which it developed broadly following the model of change initiated by its globally renown neighbouring villages of Riace and Badolato. For the past 20 years, these southern Italian villages which have embraced refugee resettlement programmes have experienced economic growth and expansion, as the local municipalities implemented long-term goals with their new migrant communities. Working together, migrants and local Italians have rejuvenated these former 'ghost towns' together – which have now become global symbols of internationalism, humanitarian aid and solidarity.

In this context around 100 migrants arrived in recent years, renovating abandoned houses in collaboration with local building masters, and joining the cooperative. The presence of these migrants has revitalized the town, leading to a modest return of activity. A restaurant, currently the only one, has opened and the primary school, which for 20 years had only one class, now has four. The primary goal is to counteract the outflow of inhabitants seeking economic opportunities elsewhere by creating local jobs. Today about 150 migrants, from Afghanistan, Morocco, Tunisia, Libya, South Sudan, Pakistan, Bangladesh, Nigeria and Syria, live in Camini, making it one of the municipalities on the peninsula with the highest rate of immigrants among its population. Their presence has sparked an outpouring of solidarity from organizations that have injected funds into the village coffers. The cooperative employs around 50 refugees who operate several workshops, including ceramics, woodworking, clothing creation and art. Some of the restored houses have been converted into bed and breakfasts, offering accommodation to visitors from all over the world who come to Camini for the inspiration and evidence of what is possible, even in one of the poorest regions of Italy.

Paraloup, Piemonte, Italy

At an elevation of 1362 m, Paraloup stands as the highest hamlet in the municipality of Rittana, situated in the Stura valley in Cuneo, Piedmont, Italy. It serves as a cultural innovation laboratory, promoting a sustainable return to mountain living by offering recreational, social and educational services, as well as activities in slow tourism and agro pastoralism. Having welcomed shepherds and flocks for centuries, Paraloup took on a new role between 1943 and 1944, during the Italian resistance. It became a laboratory for democracy for the *Italia Libera* (in English 'Free Italy') partisan band, with notable intellectual Italian figures as the writer Nuto Revelli and the lawyer Duccio Galimberti. In 2008, the Fondazione Nuto Revelli – named after the Cuneo-based alpine soldier, partisan, and writer initiated the restoration of Paraloup after a prolonged period of depopulation. The aim was to bring life and work back into the mountains, regenerating a place rich in historical knowledge and gratitude for the generations that inhabited it. Paraloup bears witness to memories, particularly those of the Resistance and the peasant world, using both architectural language and the Nuto Revelli Archive to let the stories speak. Today, Borgata Paraloup[3] has evolved into a regenerated cultural center, acting as a local hub for landscape preservation, a space of freedom and social innovation, inviting all to a shared experience of inclusiveness, solidarity and sustainability. In the etymology of the name *Paraloup,* meaning 'protected from wolves' according to local tradition, the project valued the hamlet's vocation and purpose: walking on partisan paths, living and producing in the mountains, doing business in the hamlet and fighting for just causes.

Svilengrad, Bulgaria

Situated at the crossroads of Bulgaria, Greece and Turkey, Svilengrad boasts a rich and multifaceted history shaped by its favorable natural environment and strategic geographical location. From ancient times to the present day, this dynamic region has served as a nexus of cultural exchange and economic activity, leaving an indelible mark on its landscape and inhabitants. Historically, Svilengrad has been a site of continuous human habitation, with evidence of Thracian settlements dating back to the first millennium BC. However, it was during the Ottoman period that Svilengrad truly flourished, with the construction of an impressive stone bridge over the Maritsa River in 1529 by order of Mustafa Pasha, an Ottoman general. This bridge became a focal point for trade and commerce, featuring amenities such as a caravanserai, mosque, bazaar and hammam. Despite its location along the Silk Road, an ancient trade network spanning Eurasia, Svilengrad only acquired its present name in 1913, reflecting its thriving silk farming industry. By the early 1930s, Svilengrad emerged as a leading silk producer, contributing significantly to Bulgaria's silk output. In the 20th century, Svilengrad found itself on the border of the Iron Curtain during the Cold War, shaping its sociopolitical landscape. Nowadays, the major businesses in the area are the wine industry and gambling. Since gambling is illegal in Turkey and highly restricted in Greece, visitors from both countries come over the border to Svilengrad to 'gamble all weekend, drink cheap rakia, and get their bodies rubbed with essential oils' as the Bulgarian writer Kapka Kassabova describes Svilengrad in her work *Border* (2017: 124), referring to a widespread gambling industry that is using and abusing the bordering position of Svilengrad in recent decades.

The 'Rural Migrantour' project, implemented by LAG Svilengrad,[4] placed special emphasis on the Muslim minority of Bulgarian Turks, acknowledging their historical significance and cultural contributions to the region. Descendants of both Turkish settlers and Bulgarian converts to Islam, Bulgarian Turks have played a vital role in shaping the diverse tapestry of Svilengrad's heritage. However, the community faced significant challenges during the 'Revival Process' of the 1980s, a state-sponsored assimilation policy aimed at erasing their cultural and religious identity. Forced to adopt Slavic names and renounce their Turkish heritage, many Bulgarian Turks endured persecution and discrimination during this period. Despite these hardships, the community has persevered, reclaiming their rights and asserting their cultural heritage in recent years. Today, there is greater recognition of the Bulgarian Turks' contributions to the region's cultural diversity, with initiatives aimed at preserving their traditions and promoting intercultural dialogue.

Through the 'Rural Migrantour' project, the voices and experiences of Bulgarian Turks were amplified, shedding light on their resilience in the face of adversity. By celebrating their heritage and promoting cross-cultural understanding, the project contributed to fostering a more inclusive and cohesive community in Svilengrad.

Kythera, Greece

While the island of Kythera in Greece may not be at the forefront of current discourse on contemporary refugee challenges, the island has undeniably been influenced by migration throughout its rich history. Leveraging the existing network of walking trails established by the local project consortium partner, a social cooperative *Paths of Greece*[5] and the welcoming ambiance of Kythera, the *Rural Migrantur* project aimed at exploring and highlighting the profound impact of migration on this picturesque island. Strategically positioned as a maritime crossroads connecting the Greek mainland and Crete, Kythira has served as a nexus for merchants, sailors, and conquerors throughout history, up until the mid-19th century. This intersection of cultures has bestowed upon the island a rich tapestry of influences, manifested in its architectural diversity – an exquisite fusion of traditional, Aegean, and Venetian elements. Beyond its physical structures, Kythira's customs and traditions bear witness to centuries of harmonious coexistence between Greek and Venetian cultures, weaving a unique narrative that echoes through time. The project focus extended to emigration from the small Greek island of Kythera, where the population of Kytherian origin in Australia now exceeds 20 times its current inhabitants. Male emigration to Australia, just before and after the Second World War, would not have been possible if women (wives and daughters) had not taken on the roles and activities of the departing husband, son, or brother. This shift in women's roles brought about significant changes in the management and sustenance of households, as well as in attitudes toward the micro-social environment of the village and the island at large.

Kostanjevica na Krki, Slovenia

The Gorjanci/Žumberak Mountain lies on the border between Croatia and Slovenia, encompassing a broad karst landscape of old-growth forests, caves, and species-rich grasslands. In 1530, the immigration of Uskoks, primarily of Orthodox religion, began and lasted until the 17th century. The Žumberak Mountains were then part of the Austrian Military Frontier, established by the Habsburgs as a defensive buffer between their empire and the Ottoman Empire. When the Military Frontier was disbanded in the late 19th century, mass emigration ensued. Today, the region faces continuous depopulation, an aging local population, abandonment of rural lifestyles, emigration and inadequate or poor infrastructure. The mountain, historically a crossroads with ever-changing boundaries, hosts numerous roads, tracks and paths shaped by various actors throughout history, contributing to a multi-layered landscape of past and present. Covered with woods, the mountain hides many hidden paths and tracks, once used by freight carriers, traders, and nowadays by hikers, bikers, off-road enthusiasts, hunters, and even people smugglers who inherited these paths from horse smugglers. Despite their historical significance, the Uskok communities remain enigmatic, providing an intriguing space for exploration and research. The story of the 'Uskoks' underscores the multi-faceted, multi-ethnic, and multi-cultural history embedded in rural landscapes, inviting scholars to delve into the rich tapestry of Slovenia and Croatia's past.

Within this context, Terra Vera NGO[6] plays a crucial role in fostering cross-border connections and promoting awareness of shared European heritage and values. Through initiatives focused on intercultural solidarity, migrant support and the development of intercultural interpretative tourism, it strives to promote equal rights, inclusion, and environmental well-being. As part of the European network Migrantour, it contributes to a broader effort to establish intercultural urban walks in cities across Europe, highlighting the interconnectedness of diverse communities and their histories.

Training Cross-Border Guides

The *Rural Migrantour* project focused on inclusive and participatory training for cross-border guides, with a particular emphasis on engaging young individuals from minority/migrant communities residing in peripheral/border regions. The methodology was designed to establish a supportive space for participants, considering their personal dynamics and experiences. Participants were encouraged to explore the connections between their cultural backgrounds and the cultural and historical context of the region. The training facilitated continuous knowledge development and provided a foundation for independent guidance. It was designed holistically, considering participants' emotional well-being, and addressing initial insecurities and anxieties. By creating a secure space, the project aimed to build participants' self-confidence, enhancing their ability to deliver compelling interpretations of intercultural heritage during guided walks and events.

The training for cross-border guides established a platform for exploring and sharing the historical and cultural significance of the region. The project employed a comprehensive and experiential approach, inviting participants to engage with their environment holistically, using sensory and personal methods. Through the utilization of senso-motor approaches, mindfulness practices and creative media, participants were encouraged to connect external maps of the borderland landscape with their internal experiential maps. This immersive and introspective process enhances their understanding, appreciation, and connection with the surrounding environment, fostering a deeper sense of engagement and connection with the spaces they explore. Participants had the opportunity to engage in in-depth discussions and acquire expert knowledge on various topics related to the impact of borders, peripheries, and migration on the cultural heritage of the environment. These topics included historical patterns of migration and settlement, the influence of migration on everyday life, the role of women in cross-border migration, the history of minority communities, and the evolution of borders and attitudes toward them in the 20th and 21st centuries. By immersing themselves in co-creating guided tours and learning in situ, participants developed a profound and intimate connection with the space and environments they explored, resulting in a more holistic and meaningful engagement with margins/borders and their historical significance for meaningful future scenarios.

The Training Process: Combining Different Tools and Techniques

Oral history research

Through an exploration and examination of oral testimonies from individuals residing in the peripheries/borderlands – both in the past and present – the project delved into their diverse roles and activities on cross-border walking paths. These paths, once vital for the sustainable economic and social livelihoods of locals, have been abandoned due to the recent establishment of the border. The project engaged with border residents and collected oral testimonies. This methodology allowed for the documentation and preservation of personal experiences, memories, and narratives related to the borderlands. By giving voice to individuals directly affected by border policies and historical events, the project aimed to provide a comprehensive understanding of the impact of borders on people's lives.

Counter-mapping

Employing counter-mapping which relies on reflexive cartography, the project uncovered diverse underlying realities of the borderlands, related to migration, minority and gender issues. This approach challenged dominant narratives and power structures by mapping the presence and contributions of minorities, migrants and women in the construction of peripheries and transgression of borders. It allowed for a visual representation of women's roles and activities, creating a register of landmarks on an interactive map to highlight their significance.

Co-creation

The methodological approach of the project emphasized a participatory and collaborative approach by involving participants of the training for intercultural guides in the research, development and implementation processes. By engaging the local communities and individuals affected by border policies, the project ensured that their voices are heard, their needs are considered, and their perspectives are included in the project outcomes. This approach promoted empowerment, inclusivity, and the co-ownership of the project's results.

Multimedia production

To reach a wider audience and make the borderlands more accessible, multimedia tools were employed such as video capsules, podcast interviews and autobiographical testimonies. These mediums enabled the project to present the collected oral histories and stories in engaging and accessible formats. Multimedia production also facilitated the creation of interactive multimedia story maps, allowing for the digital preservation and dissemination of the heritage of the borderlands.

Project Outcomes: Cross-Borders Guides and Intercultural Walks

In order to provide a comprehensive perspective from migrant and/or minority communities, we included testimonies of migrants currently experiencing the challenges of illegally crossing these borders. Their first-hand accounts shed light on the realities and difficulties faced by those seeking a better life and the impact it has on their lives. By incorporating these testimonies, we fostered a deeper understanding of the human aspect of migration, promoting empathy and dialogue. This approach allowed us to bridge the two perspectives, creating a more comprehensive understanding of the complexities surrounding migration and border dynamics. By creating an inclusive environment, the project sought to give a voice to all residents and capture the richness of their stories and experiences. Through the processes of co-creation, the project developed hiking itineraries that promote intercultural heritage. Today, the trails offer opportunities to strengthen the understanding and potential of inclusive tourism in marginal regions, involving young individuals in cultural tourism as interpretative guides and product providers, thereby strengthening their economic and emancipatory role in the community and beyond.

As a part of project dissemination, the newly trained intercultural companions offered guided walks aimed at selected targets (schools, local authorities, non-profit organisations, media, regional tourist service providers) and general public. During the walk, the guides share stories related to the local environment and their experiences, focusing on the impact of diverse groups and individuals associated with borders, migration, and the local environment in the past and present. The program includes multisensory learning experiences: taste (e.g. food and drink), smell (e.g. specific smells relating to the themes of particular experiences), vision (e.g. photographs, films), touch (e.g. objects, etc.), and hearing (e.g. music). The guides emphasize the values of cross-border cultural tourism and diverse views/interpretations of shared cultural heritage. An interactive multimedia map of peripheral/border itineraries has been developed, promoting and disseminating the project results while reaching a wider audience, including potential participants in guided tours and potential new guides. The interactive maps are now part of a new generation of learning tools based on participatory methods. By utilizing new technologies for online storytelling, the project aims to increase the interest of target groups and the general public in alternative, marginalized and overlooked cultural heritage and related (auto)biographies, not only in Europe but also beyond. This contribution seeks to reduce intolerance and foster acceptance of cultural diversity.

Conclusions: Inclusive Cross-Border Cultural Tourism as a Utopian Forecast

As we contemplate the future trajectories of tourism, informed by the insights gained from the Rural Migrantour project, a nuanced understanding of the dynamics of 'otherness' and its impact on tourism narratives emerges. Grounded in the project's ethos of inclusivity and cultural exchange, we are compelled to reimagine tourism futures within the context of border areas, where the complexities of identity, belonging, and socioeconomic disparity intersect. The Rural Migrantour project serves as a seminal case study in the deconstruction of traditional tourism paradigms, challenging the prevailing narratives of exoticism and othering that have long characterized tourist encounters. By foregrounding the voices and experiences of marginalized communities residing in border regions, this initiative has engendered a transformative dialogue that transcends cultural divides and fosters mutual understanding. In doing so, it lays the foundation for a tourism future predicated on empathy, reciprocity and ethical engagement in rural context – as opposed to similar initiatives that are usually focused on urban settings.

Envisioning the future of tourism in border areas necessitates a multifaceted appraisal of its economic and social development. On one hand, the project's emphasis on responsible tourism and community empowerment holds the promise of socioeconomic and cultural revitalization. By leveraging the rich tapestry of stories and heritage embedded within these locales, tourism becomes a catalyst for economic resilience and social cohesion. Moreover, the engagement of local youth from minority and migrant backgrounds as guides not only enhances their economic prospects but also facilitates intergenerational knowledge transmission and cultural preservation. However, this vision of tourism futures is not without its attendant risks and constraints. The inherent complexities of border regions, marked by historical legacies of conflict, displacement, and geopolitical tensions, pose significant challenges to sustainable tourism development. Moreover, the commodification of culture and the spectre of exploitation loom large, necessitating a nuanced approach to tourism governance and regulation.

By framing inclusive cross-border cultural tourism as a utopia scenario, we align with the ontological typology of futures studies proposed by Bergman *et al.* (2010). In their exploration of truth claims and explanatory claims, they identify the 'utopia forecast' as a category of futures thinking that envisions radical yet plausible societal transformations. In this sense, initiatives like Rural Migrantour represent a tangible enactment of utopian tourism – not as an unattainable fantasy, but as a guiding vision for more equitable and inclusive tourism practices. By fostering deep cultural exchanges, reconfiguring notions of borders, and centering marginalized voices, cross-border tourism emerges as a speculative yet actionable model that challenges exclusionary policies and entrenched power dynamics. Within this framework, tourism functions not merely as an economic driver but as a transformative force capable of reimagining mobility, identity, and belonging beyond national and cultural divides.

In conclusion, the Rural Migrantour project offers a compelling vision of tourism futures that transcends the boundaries of otherness and fosters genuine dialogue and exchange. As we navigate the intricate terrain of border area tourism, let us remain dedicated to our commitment to ethical engagement, cultural sensitivity, and equitable development. Only through a holistic approach that acknowledges the complexities of identity and heritage can we realize the transformative potential of tourism as a catalyst for positive change.

Notes

(1) https://sites.google.com/view/ruralmigrantour/home
(2) https://eurocoopcamini.com/
(3) https://paraloup.it/
(4) https://migsvilengrad.org/
(5) www.pathsofgreece.gr/en/
(6) www.terra-vera.org

References

Bergman, A., Karlsson, J.C. and Axelsson, J. (2010) Truths claims and explanatory claims – an ontological typology of futures studies. *Futures* 42 (8), 857–865.

Dikomitis, L. (2005) Three readings of a border: Greek Cypriots crossing the green line in Cyprus. *Anthropology Today* 21 (5), 7–12.

Kassabova, K. (2017) *Border: A Journey to the Edge of Europe*. Graywolf Press.
Lundén, T. (ed.) (2006) *Crossing the Border: Boundary Relations in a Changing Europe*. Gondolin.
Navaro-Yashin, Y. (2012) *The Make-Believe Space: Affective Geography in a Postwar Polity*. Duke University Press.

13 Digital Worlds for Blue Tourism: The Case of Extended Realities in Tourism Futures

Michal Lovecky and Ninja Mueller

Introduction

Tourism involves a departure from daily routines, engaging senses with stimuli contrasting the mundane (Urry, 2015). In this context, Blue Tourism, defined as all tourism activities related to oceans, seas, and coasts (Tonazzini *et al.*, 2019), offers a distinctive marine environment vastly different from terrestrial settings. Blue Tourism, a sector within the expansive domain of travel encompassing Coastal, Marine and Maritime Tourism (CMMT), merges the marine world with sustainable practices, seeking to protect the diverse ecosystems beneath and above the waves (UN-DESA, 2016).

However, the rapid development of Blue Tourism poses specific challenges for the marine ecosystems and the coastal communities, such as waste generation and pollution of the coastal environment (e.g. Beeharry *et al.*, 2017; Silva & Ghilardi-Lopes, 2012), destruction of flora and fauna (e.g. Barker & Roberts, 2004; Mendez *et al.*, 2017) and disruption of community and wildlife behaviour (e.g. Brebbia, 2008; De Luca *et al.*, 2020). Poorly managed Blue Tourism can lead to marine ecosystem exploitation and destruction, degrading the preconditions for attractive and beneficial CMMT (Bennett *et al.*, 2021; Tonazzini *et al.*, 2019). Subsequently, the availability of ecosystem services for the livelihood of local communities alongside cultural values connected to the marine environment are reduced (Leposa, 2020). Furthermore, like other kinds of tourism, Blue Tourism can foster the dispossession of lands and waters and exclude locals from decision-making processes (Bennett *et al.*, 2021). Especially marginalized groups such as indigenous people or women are vulnerable to exploitation in Blue Tourism (Bennett *et al.*, 2021; Fletcher *et al.*, 2020). At the same time, Blue Tourism hotspots, mainly located in the Global South, often depend on the economic value of Blue Tourism and have to balance its negative impacts and dependencies with its economic benefits (Brown & Hall, 2008; Rogerson & Saarinen, 2018).

Tourism reflects the complexities of modern society, where technology increasingly influences daily interactions (UNWTO, 2019; Walton, 2023). In the Blue Tourism sector, emerging technologies such as immersive media and digital worlds – often regarded as the next evolution of the internet – signify a revolutionary advancement in digital interaction and touristic experiences (Cheng *et al.*, 2022; Lee *et al.*, 2021). Utilizing extended reality technologies, which include augmented reality (AR; a technology that superimposes digital information onto the physical environment), virtual reality (VR; a fully immersive digital experience that replaces the physical surroundings), mixed reality (an integration of real and virtual worlds to produce new environments where physical and digital objects co-exist), and 3D graphics, digital worlds facilitate real-time interactions in ways that are not possible in the physical realm. As a platform, digital worlds offer unparalleled opportunities for sectors ranging from marketing and tourism to healthcare and education, transforming how businesses can engage with consumers and stakeholders (Koohang *et al.*, 2023). The incorporation of these technologies in tourism provides new, simulated experiences that can deliver an enhanced sense of presence, interaction, and agency (Beck *et al.*, 2019; tom Dieck *et al.*, 2018), effectively diminishing the boundary between reality and virtuality.

Immersive media in Blue Tourism offers an opportunity to both bridge and highlight the gap between humans and the marine environment, providing a space for a deeper understanding and appreciation of the marine other without objectifying or commodifying it. The intersection of Blue Tourism with immersive media and metaverse hints at a future where CMMT experiences transcend physical boundaries. This fusion can democratize ocean access, making it accessible to all, irrespective of socio-economic status or physical capability. Furthermore, the digital immersion into the marine environment enables a shift in perspectives beyond the anthropocentric lens. Marine animals are subject to othering on a more extensive scale than terrestrial non-human-others, rendering the ocean and its inhabitants an almost alien environment for many (Conley, 2008; Fauville *et al.*, 2019). However, as immersive media like VR are increasingly researched, the power of evoking empathy for others and bridging alienation is investigated and shows promising results (e.g. Ahn *et al.*, 2016; Barbot & Kaufman, 2020; Schutte & Stilinović, 2017). To name just two examples: Ahn *et al.* (2016) conducted research on the increased interconnection with the natural environment of participants after embodying non-human animals in VR; Schutte and Stilinović (2017) found a higher level of empathy of participants for characters they joined in VR than watched on a flat screen. Though promising, the fusion of Blue Tourism and immersive media also raises critical questions that shall be addressed in this contribution. To what extent can future virtual experiences deepen our connection to the ocean, or conversely, might they inadvertently trivialize these ecosystems? Additionally, what might be the impact of virtual travel on the socioeconomic dimensions of coastal communities in the future? This chapter aims to address and discuss these questions, by interweaving its critical reflection with Bergman *et al.*'s (2010) ontological typologies of futures studies to present and discuss a utopian and a dystopian scenario, which helps to draw conclusions for the ethical use of extended realities in tourism futures.

State of the Art

The integration of immersive media – VR, augmented reality (AR), and 360-degree video – has been transformative for socio-technological relationships within contemporary tourism (van Nuenen & Scarles, 2021). Immersive media is defined as follows by Kaplan-Rakowski and Meseberg 'as media that stimulate physical senses to the point where we experience psychological immersion, also referred to as telepresence. [Immersive media] have the capability to make users so involved in the filmed, photographed, synthetic, or mixed environment that they have an impression that it is real and that they are present in this environment (so-called being there)' (Kaplan-Rakowski & Meseberg, 2019: 144). This evolution has been characterized by significant advancements in both hardware and software over the past two decades. Pioneering projects like the digital worlds of *Second Life* (Second Life, 2024), offering various digital experiences and life simulations, including virtual recreations of real-world locations; *Rome Reborn*, an international initiative launched in the mid-1990s by the UCLA Cultural Virtual Reality Laboratory, providing a comprehensive digital reconstruction of ancient Rome and merging educational objectives with virtual tourism experiences (Flyover Zone, 2018); and virtual tours of UNESCO World Heritage Sites, are examples of early ventures that paved the way for more sophisticated applications in the tourism industry.

The introduction of experiences such as Thomas Cook's 'Try Before You Fly' marked the shift in the use of VR in the tourism industry. Deployed as a 'try before you buy' mechanism, VR allowed travellers to experience a holiday in New York before purchasing their trips, boosting agency's revenue from VR-promoted New York trips by 190 percent (Visualise, n.d.). Around the same time, the *Sheepview 360°* campaign by *Visit Faroe Islands* creatively used an animal-borne 360° video to promote the Faroe Islands. This initiative involved equipping sheep with 360° cameras to capture the island's landscapes from their perspective, demonstrating the potential of immersive media in tourism PR and destination marketing, while directly engaging with non-human actors (Visit Faroe Islands, 2017). This and similar services provided a glimpse into the potential of immersive technologies in enhancing guest experiences.

In the realm of Blue Tourism, immersive media have been instrumental in bridging the gap between humans and the marine environment. Projects like *theBlu* by Wevr Inc. offer immersive VR experiences of the ocean's habitats, bringing users face-to-face with awe-inspiring marine species (Wevr, n.d.). Similarly, *Titanic VR* by Immersive VR Education Ltd. combines a diving simulator with an interactive story, bringing the historical tragedy of the RMS Titanic to life (Steam, 2018). This experience not only educates users about the event but also provides an evocative understanding of the accident.

In the context of conservation and education, projects of the immersive media production studio Cyan Planet like *How is the Water*, where users experience ocean life from a dolphin's perspective, and the 'Seychelles Coral Reef Restoration' 360° video, highlight the challenges marine ecosystems are facing (Cyan Planet, 2022). Further research demonstrates the potential of VR and 360° video in fostering empathy for marine life and to inspire or support real-world conservation efforts (Ahn *et al.*, 2016; Barbot & Kaufman, 2020; Blythe *et al.*, 2021).

The adoption of immersive technologies such as VR and AR in Blue Tourism could initiate a shift towards more inclusive and educational travel experiences (Buhalis *et al.*, 2023). This shift would extend the reach of tourism, allowing a broader audience to experience destinations virtually, breaking down traditional barriers of travel. Such technological advancements play an important role in enriching educational content and in sensitizing virtual tourists to ecological and cultural heritage, thus contributing to a more sustainable and responsible tourism sector (Fan *et al.*, 2022; Guttentag, 2010; Loureiro *et al.*, 2020). Additionally, these technologies are reshaping marketing strategies within the industry, offering interactive previews that can impact travel planning, purchase decisions and enhance customer engagement.

Exploring Scenarios: The Future of Blue Tourism in a Metaverse-Mediated World

Two possible scenarios for the future development of Virtual Blue Tourism are explored (Bergman *et al.*, 2010). The terms 'digital worlds' and 'metaverse' are used interchangeably. The first scenario outlines an optimistic, near-to-utopian vision, anticipating how the metaverse could support inclusivity and bridge the concept of otherness. Conversely, the second scenario paints a pessimistic, near-to-dystopian picture, illustrating how these technologies, if not responsibly implemented, could reinforce processes of othering and inequity. Furthermore, reflecting on the future integration of immersive media and digital worlds into Blue Tourism, it is important to consider a multitude of perspectives shaped by different beliefs in sustainable development and socio-natural relations conditioned by ocean's dimensionality and fluidity. This ranges from those who lean towards a techno-centric approach, emphasizing technological solutions, to those who advocate for an eco-centric approach, prioritizing environmental considerations. In this context, a stance of environmental pluralism is adopted, recognizing a diversity of perspectives, values and approaches (Curry, 2003; Isacs *et al.*, 2023).

From this vantage point, a technological future is envisioned where the metaverse is on the cusp of fulfilling its predicted role as the next evolutionary stage of the internet (Cheng *et al.*, 2022; Lee *et al.*, 2021). This future metaverse is accessible through a wide range of affordable devices offering varying degrees of immersion and agency. In this envisioned future, creating digital worlds through dedicated AI-powered tools and templates is relatively easy and affordable, similarly to today's creation of websites. However, the cost can escalate significantly with higher demands for uniqueness, data privacy, security and other advanced features. A robust economy has emerged within the metaverse, featuring virtual equivalents of traditional occupations, including many new ones, such as virtual architects, virtual real estate developers, digital identity consultants, virtual tourist guides, merchants, and more. These professionals sell various digital goods, services, and experiences, much like their counterparts in the physical world.

In this future, the metaverse, particularly its ocean-related instances, is connected to comprehensive digital twins. The digital twin of the ocean is a virtual replica of ocean ecosystems powered by advanced AI models, billions of remote ocean sensors, and other data sources (Chen *et al.*, 2023; European Commission, 2023). As a result, any single digital world within the metaverse can not only narrate any story its creator chooses but can also replicate any marine or coastal environment with high accuracy, reflecting either immediate, historical, or modelled future conditions. Given this brief introduction into the technological landscape, the possible impacts of such developments towards a Virtual Blue Tourism shall be considered in the following. Those impacts are explicated on the individual, local, and global level of tourism.

Digital worlds for inclusivity and accessibility: A utopian scenario

Drawing on Bergman *et al.* (2010) ontological typologies of tourism studies, a utopian future scenario can be envisioned with regards to the integration of the metaverse into Blue Tourism bringing a new era of inclusivity and accessibility. This technological evolution promises to make marine and coastal experiences universally accessible, overcoming physical and geographical limitations, including the challenges of digital divide (Lythreatis *et al.*, 2022). Virtual tourism becomes a powerful tool that breaks down socioeconomic barriers to global travel and allows people from all walks of life to explore and interact with various aspects of Blue Tourism, such as remote underwater spaces and culturally rich coastal communities, from anywhere in the world.

For the individual, virtual tourism evokes effects similar as conventional tourism, such as better understanding other cultures, organisms, or ecosystems (Fan *et al.*, 2022; Verma *et al.*, 2022), however, it is enriched by the advancements of digital worlds. The potential for re-experiencing and re-connecting is expanded, allowing individuals to re-engage with people, cultures, and places, and to share these experiences with any travel companion, no matter where they are. Digital worlds serve as bridges that reduce the otherness perceived in distant and unfamiliar environments and beings and provide a platform for global (ocean) stewardship, linking the digital experiences with real-world counterparts (European Commission, 2024).

The capacity of virtual reality and digital worlds for creating immersive and emotionally engaging experiences holds the potential to transform our perception of nature (Schutte & Stilinović, 2017; Shin, 2018; Trevena *et al.*, 2024; Ventura *et al.*, 2020). These technologically mediated transformative experiences could redefine humanity's relationship with the marine environment, elevating the human from the mere observer to an active participant in a shared ecosystem. The possibility of experiencing the world through the perspectives of non-human entities, such as marine animals, as already exemplified by Cyan Planet's How is the Water experience (Cyan Planet, 2021), could deepen the ecological consciousness and empathy, empowering collective global action (Ahn *et al.*, 2016; Blythe *et al.*, 2021; Herrera *et al.*, 2018).

The digital twin of the ocean, a core component of this virtual realm, presents an opportunity to foster a deeper understanding and connection with the marine environment. Within this future technological landscape lies the ability to create hyper-realistic, interactive digital replicas of environments. Among other things, this enables a form of digital conservation, preserving or bringing back to life organisms and ecosystems of the marine world that might otherwise be lost to time and environmental change (Digital Life, 2017). Especially in a time some call the Holocene extinction (Braje & Erlandson, 2013; Turvey & Crees, 2019), the digital preservation of vulnerable or extinct species through 3D modelling and animation could offer educational awareness about what is absent that conventional tourism is probably not able to convey. While it is inherent that human perspectives will always carry a degree of anthropocentrism, the digital twins offer a unique opportunity to challenge and potentially recalibrate this viewpoint.

Apart from the virtual blue tourist, the geographic location of virtually travelled places is also positively impacted by its digital twin: Reducing physical visits alleviates stress from marine ecosystems, as digital worlds deliver close-to-real world experiences (Parcu *et al.*, 2023; Polishchuk *et al.*, 2023). The created digital spaces facilitate collective action by providing stakeholders and stewards a platform to witness, engage, and experiment with various conservation actions (Games for Change, 2022; Humble Bundle, 2022). The opportunity to confront a virtual tourist with the ideal versus current state of marine ecosystems, for example a healthy coral reef teeming with life versus the same coral reef struggling with plastic pollution, overfishing, or coral bleaching, provides a unique experience that could encourage a feeling of shared responsibility and possibly empower more effective conservation efforts through Virtual Blue Tourism. Local actors can, for example, host virtual morning walks improving the virtual tourists' well-being, and fundraise globally to support activities such as coastal protection projects or a beach cleanup (Frost *et al.*, 2022).

Furthermore, the metaverse supports the development of decentralized governance structures, redefining regulatory frameworks to enhance democratic decision-making and community engagement. These frameworks facilitate more equitable and effective management of both digital and real-world resources. In other words, it provides a unique platform allowing for inclusive and equitable participation of a diverse array of actors. Especially for stakeholders in tourism, often small and medium local businesses, the digital world enables them to participate in processes conditioning their activities, and aids in overcoming their current challenges to participation such as resource constraints and inadequate inclusivity (Frazão Santos *et al.*, 2021).

The integration of the metaverse creates a thriving economic landscape for local businesses. Transitioning to virtual tourism, these businesses harness the global market, linking their real-world offerings with digital counterparts, creating a new revenue stream that supplements traditional income, for example, a virtual tour where participants purchase local souvenirs, which are then physically delivered. On a global scale, the access to travel the ocean and coastal areas is democratized and not dependent on a virtual tourist's wealth. Travelling distant places to learn about cultures and ecosystems is affordable for everybody. Coastal and indigenous communities find their voices amplified within these digital worlds, gaining greater control and representation. This empowerment helps preserve cultural and historical authenticity in virtual spaces.

Digital worlds for *othering* processes and commodification: A dystopian scenario

Drawing on Bergman *et al.* (2010) ontological typologies of tourism studies, it can be envisioned a dystopian scenario in which the metaverse integration exacerbates existing socioeconomic divisions and empowers commodification of nature. Digital worlds, crafted for quick entertainment, detach individuals from the physical world, offering a form of escapism and reinforcing a distorted perception of the true state of nature.

In this scenario, the digital becomes an idealized reflection, dividing the virtual tourists, and deepening the alienation from nature needing protection. Adding to this vision a future dimension, where climate change mitigation and adaptation actions are still questioned, the metaverse becomes a tool that dilutes important action and creates a space for radicalization of worldviews which further reinforces othering of humans and non-humans. A new form of *solastalgia* (Albrecht *et al.*, 2007) impacts metaverse tourists, as they are confronted with the dichotomy of their virtual idealized digital worlds and the reality of the actual physical state of nature, impacting their mental health and well-being.

The growing disconnection between people and nature, where nature is reduced to an object of entertainment for human pleasure, further fuels environmental degradation (Büscher, 2022). Individuals, without opportunities to cultivate a genuine connection with nature, fail to grasp the impact of their actions, which leads to increased pollution and continued destruction of coastal and marine ecosystems. As the marine environment degrades and baselines shift further, nature, once captivating, loses its qualities and makes it harder for the public to develop an understanding and appreciation of its vital role in our well-being and survival. This creates a cycle of neglect, supported by virtual detachment from the real-world consequences. Early adopters and resource-rich service providers in the metaverse have a competitive edge and use this advantage to dominate the market and exert pressure on those who are either directly excluded from the digital spaces or need to offer unique value propositions that are hard to copy by the dominating players to attract virtual tourists. Such dynamics exacerbate economic disparities, favouring those who can create appealing virtual experiences for global audiences. In this future, platforms hosting these digital worlds capture a significant share of the economic value, impacting local profit generation, and in the name of 'disruption', create skewed economic structures.

Since equitable access to these emerging technologies is not the norm, the democratization of virtual tourism experiences allowing diverse global communities to engage with and benefit from these platforms is only a mirage, impacting marginalized cultures, underrepresented groups and indigenous people's rights. By design, the metaverse offers a platform for diverse groups to share their narratives, but as seen in the evolution of social networks, having a voice doesn't guarantee being heard (Gillespie, 2018; Myers West, 2018; Napoli, 2019). The platforms managing these digital worlds face challenges akin to those of current social networks, which often polarize discourse and favor sensational, click-worthy content (Guerra *et al.*, 2013). Such biased dynamics undermine the authenticity and impact of narratives shared by indigenous and local communities, further complicating the interplay between virtual tourism, cultural representation and rights.

In the dystopian scenario, the metaverse and digital ocean twin infrastructure exert tremendous pressure on environmental sustainability through high energy consumption and increased e-waste. The infrastructure required to support a widespread, daily-used metaverse – including cloud services, access devices, storage solutions, and the energy-intensive processes of rendering and simulation – presents significant environmental challenges. This infrastructure, designed to replicate marine environments, becomes a paradox as it contributes significantly to environmental degradation, indirectly impacting Blue Tourism dependent on a healthy ocean. Simultaneously, the reliability of technology escalates to a point where automated systems dictate critical responses to environmental alerts. For instance, a cyber-attack targeting these systems could mistakenly trigger a tsunami warning, leading to unnecessary evacuations and economic disruption for coastal communities reliant on tourism. Moreover, a variety of decision-making processes such as destination selection becomes overly dependent on digital twins and predictive models, which, despite their advanced algorithms, still carry the inherent uncertainty of any future-oriented prediction. This dependency cultivates a false sense of security and overpromises on tourism experiences.

Instead of a global accessibility to Virtual Blue Tourism, high-quality virtual experiences become luxury commodities, be it due to requirements on personal hardware or a walled-garden approach of service providers, privileging a selected few and creating a new form of othering, where those unable to afford or access the metaverse are marginalized from these experiences. Instead of being a tool for environmental connection and education, it further divides what is truly lived and what is experienced virtually, trivializing marine environments, and turning them into mere entertainment backdrops.

Conclusions

Being presented as a utopian and a dystopian scenario in alignment with Bergman *et al.*'s (2010) futures studies, both visions are sides of the same coin. The currently existing technologies for creating digital worlds hold the potential to develop in either of these directions, with powerful implications for humanity's understanding of space and its perception of otherness. However, the direction this technology takes depends on how influential industries use it in the future. The tourism industry and specifically Virtual Tourism hold great potential to shape the future use of immersive technologies (Buhalis *et al.*, 2023), therefore, a careful approach to the design and implementation of these virtual experiences is imperative, ensuring they complement and support rather than replace real-world engagement and action. Initiatives that bridge virtual experiences with real-world conservation efforts are gaining traction, driven by an eco-conscious society today. Platforms like *Games for Change* and various game studios are already leveraging gaming to support sustainable development, combining entertainment with educational and conservation objectives. These projects not only raise awareness but also directly contribute to conservation efforts through fundraising and community engagement (Games for Change, 2022; Humble Bundle, 2022).

Ethical considerations also play an important role in this transition. As the intersection of technology, tourism, and conservation is navigated, it is essential to craft solutions and policies that address the complex array of issues that arise. These range from data privacy and security in the digital realm to the ethical representation of peoples, cultures, spaces, marine life and ecosystems. For sustainable blue tourism, these policies and considerations need to be deeply embedded in the inner workings of future metaverse platforms. The concept of Blue Justice (Bennett *et al.*, 2021), woven into the development of these virtual tourism experiences, can be a guide towards a future where technology not only advances capabilities but also upholds ethical responsibilities. This principle can ensure that the benefits and burdens of Virtual Blue Tourism are distributed equitably, fostering a tourism landscape that is not only technologically advanced but also socially responsible and environmentally sustainable.

Acknowledging the role of the tourism industry in responsibly advancing immersive technologies, further research on utilising digital worlds for virtual tourism experiences is needed. This technological evolution brings forth many complexities. While it should strive for inclusive and sustainable tourism practices, it simultaneously poses questions regarding the authenticity of experiences and the impacts on local ecosystems and communities. Therefore, research should not focus only on the impact of virtual tourism on its 'users', the tourists, but also include a perspective on its implications for the local infrastructures. These challenges might be addressed by decentralized governance structures that actively involve all stakeholders and offer a pathway to ensuring that these digital spaces are equitable and representative. By giving diverse voices, including those of underrepresented and indigenous communities, an equal opportunity to be heard and to influence platform policies, digital worlds can become a tool for genuine engagement and cultural exchange rather than mere digital escapism.

In conclusion, the intersection of Blue Tourism and the metaverse heralds a future rich with innovative possibilities. Virtual Blue Tourism might revolutionize the engagement with blue spaces and the perception of human and non-human otherness, and thereby improve livelihoods and opportunities for coastal communities. Yet, this journey requires a balanced approach, where technological innovation is harmonized with ethical considerations, environmental stewardship, and a commitment to inclusivity and equity.

References

Ahn, S.J.G., Bostick, J., Ogle, E., Nowak, K.L., McGillicuddy, K.T. and Bailenson, J.N. (2016) Experiencing nature: Embodying animals in immersive virtual environments increases inclusion of nature in self and involvement with nature. *Journal of Computer-Mediated Communication* 21 (6), 399–419.

Albrecht, G., Sartore, G.-M., Connor, L., Higginbotham, N., Freeman, S., Kelly, B., Stain, H., Tonna, A. and Pollard, G. (2007) Solastalgia: The distress caused by environmental change. *Australasian Psychiatry* 15 (1), S95–S98.

Barbot, B. and Kaufman, J.C. (2020) What makes immersive virtual reality the ultimate empathy machine? Discerning the underlying mechanisms of change. *Computers in Human Behavior* 111, 106431.

Barker, N.H.L. and Roberts, C.M. (2004) Scuba diver behaviour and the management of diving impacts on coral reefs. *Biological Conservation* 120 (4), 481–489.

Beck, J., Rainoldi, M. and Egger, R. (2019) Virtual reality in tourism: A state-of-the-art review. *Tourism Review* 74 (3), 586–612.

Beeharry, Y.D., Bekaroo, G., Bokhoree, C., Phillips, M.R. and Jory, N. (2017) Sustaining anti-littering behavior within coastal and marine environments: Through the macro-micro level lenses. *Marine Pollution Bulletin* 119 (2), 87–99.

Bennett, N.J., Blythe, J., White, C.S. and Campero, C. (2021) Blue growth and blue justice: Ten risks and solutions for the ocean economy. *Marine Policy* 125, 104387.

Bergman, A., Karlsson J.C. and Axelsson, J. (2010) Truths claims and explanatory claims – an ontological typology of futures studies *Futures* 42 (8), 857–865.

Blythe, J., Baird, J., Bennett, N., Dale, G., Nash, K.L., Pickering, G. and Wabnitz, C.C.C. (2021) Fostering ocean empathy through future scenarios. *People and Nature* 3 (6), 1284–1296.

Braje, T.J. and Erlandson, J.M. (2013) Human acceleration of animal and plant extinctions: A Late Pleistocene, Holocene, and Anthropocene continuum. *Anthropocene* 4, 14–23.

Brebbia, C.A. (2008) *Environmental Problems in Coastal Regions VII* (Vol. 99). WIT.

Brown, F. and Hall, D. (2008) Tourism and development in the global south: The issues. *Third World Quarterly* 29 (5), 839–849.

Buhalis, D., Leung, D. and Lin, M. (2023) Metaverse as a disruptive technology revolutionising tourism management and marketing. *Tourism Management* 97, 104724.

Büscher, B. (2022) The nonhuman turn: Critical reflections on alienation, entanglement and nature under capitalism. *Dialogues in Human Geography* 12 (1), 54–73.

Chen, G., Yang, J., Huang, B., Ma, C., Tian, F., Ge, L., Xia, L. and Li, J. (2023) Toward digital twin of the ocean: From digitalization to cloning. *Intelligent Marine Technology and Systems* 1 (1), 3.

Cheng, R., Wu, N., Chen, S. and Han, B. (2022) Will metaverse be nextG internet? vision, hype, and reality. *IEEE Network* 36 (5), 197–204.

Conley, M.K. (2008) Caring for dolphins, otters, and octopuses: Speciesism in the regulation of zoos and aquariums. *Animal Law* 15, 237–264.

Curry, P. (2003) Re-thinking nature: Towards an eco-pluralism. *Environmental Values* 12 (3), 337–360.

Cyan Planet (2021) *How is the Water*. Cyan Planet. Retrieved from: https://www.cyanplanet.org/how-is-the-water

Cyan Planet (2022) *Cyan Planet | VR for Ocean Protection*. Cyan Planet. Retrieved from: https://www.cyanplanet.org

De Luca, G., Shirvani Dastgerdi, A., Francini, C. and Liberatore, G. (2020) Sustainable cultural heritage planning and management of overtourism in art cities: Lessons from atlas world heritage. *Sustainability* 12 (9), 3929.

Digital Life (2017) Digital life—creating accurate 3d models of life on earth. *Digital Life*. Retrieved from: http://digitallife3d.org

European Commission (2023) *European Digital Twin of the Ocean (European DTO). Research and Innovation*. Retrieved from: https://research-and-innovation.ec.europa.eu/funding/funding-opportunities/funding-programmes-and-open-calls/horizon-europe/eu-missions-horizon-europe/restore-our-ocean-and-waters/european-digital-twin-ocean-european-dto_en

European Commission (2024) *Virtual Worlds Fit for People. Shaping Europe's Digital Future*. Retrieved from: https://digital-strategy.ec.europa.eu/en/policies/virtual-worlds

Fan, X., Jiang, X. and Deng, N. (2022) Immersive technology: A meta-analysis of augmented/virtual reality applications and their impact on tourism experience. *Tourism Management* 91, 104534.

Fauville, G., Payne, D.L., Marrero, M.E., Lantz-Andersson, A. and Crouch, F. (eds) (2019) *Exemplary Practices in Marine Science Education*. Springer International Publishing.

Fletcher, R., Mas, I.M., Romero, A.B. and Blázquez-Salom, M. (2020) *Tourism and Degrowth: Towards a Truly Sustainable Tourism*. Routledge.

Flyover Zone (2018) *Rome Reborn—Flight over Ancient Rome. Flight Over Ancient Rome*. Retrieved from: https://www.flyoverzone.com/rome-reborn-flight-over-rome/

Frazão Santos, C., Agardy, T., Andrade, F., Crowder, L.B., Ehler, C.N. and Orbach, M.K. (2021) Major challenges in developing marine spatial planning. *Marine Policy* 132, 103248.

Frost, S., Kannis-Dymand, L., Schaffer, V., Millear, P., Allen, A., Stallman, H., Mason, J., Wood, A. and Atkinson-Nolte, J. (2022) Virtual immersion in nature and psychological well-being: A systematic literature review. *Journal of Environmental Psychology* 80, 101765.

Games for Change (2022) *Games for Change. Games for Change*. Retrieved from: https://www.gamesforchange.org/

Gillespie, T. (2018) *Custodians of the Internet: Platforms, Content Moderation, and the Hidden Decisions that Shape Social Media*. Yale University Press.

Guerra, P., Jr., W.M., Cardie, C. and Kleinberg, R. (2013) A measure of polarization on social media networks based on community boundaries. *Proceedings of the International AAAI Conference on Web and Social Media* 7 (1), Article 1.

Guttentag, D.A. (2010) Virtual reality: Applications and implications for tourism. *Tourism Management* 31 (5), 637–651.

Herrera, F., Bailenson, J., Weisz, E., Ogle, E. and Zaki, J. (2018) Building long-term empathy: A large-scale comparison of traditional and virtual reality perspective-taking. *PLOS ONE* 13 (10), e0204494.

Humble Bundle (2022) *Humble Bundle Main Page. Humble Bundle*. Retrieved from: https://www.humblebundle.com/charities

Isacs, L., Kenter, J.O., Wetterstrand, H. and Katzeff, C. (2023) What does value pluralism mean in practice? An empirical demonstration from a deliberative valuation. *People and Nature* 5 (2), 384–402.

Kaplan-Rakowski, R. and Meseberg, K. (2019) Immersive media and their future. In R.M. Branch, H. Lee and S.S. Tseng (eds) *Educational Media and Technology Yearbook* (Vol. 42, pp. 143–153). Springer International Publishing.

Koohang, A., Nord, J.H., Ooi, K.-B., Tan, G.W.-H., Al-Emran, M., Aw, E.C.-X., Baabdullah, A.M., Buhalis, D., Cham, T.-H., Dennis, C., Dutot, V., Dwivedi, Y.K., Hughes, L., Mogaji, E., Pandey, N., Phau, I., Raman, R., Sharma, A., Sigala, M. … Wong, L.-W. (2023) Shaping the metaverse into reality: A holistic multidisciplinary understanding of opportunities, challenges, and avenues for future investigation. *Journal of Computer Information Systems* 63 (3), 735–765.

Lee, L.-H., Braud, T., Zhou, P., Wang, L., Xu, D., Lin, Z., Kumar, A., Bermejo, C. and Hui, P. (2021) *All One Needs to Know about Metaverse: A Complete Survey on Technological Singularity, Virtual Ecosystem, and Research Agenda*. arXiv.Org, abs (2110.05352). Retrieved from: https://www.semanticscholar.org/paper/All-One-Needs-to-Know-about-Metaverse%3A-A-Complete-Lee-Braud/b00cc531f4872dcff8577fc238f70a088eef3b56

Leposa, N. (2020) Problematic blue growth: A thematic synthesis of social sustainability problems related to growth in the marine and coastal tourism. *Sustainability Science* 15 (4), 1233–1244.

Loureiro, S.M.C., Guerreiro, J. and Ali, F. (2020) 20 years of research on virtual reality and augmented reality in tourism context: A text-mining approach. *Tourism Management* 77, 104028.

Lythreatis, S., Singh, S.K. and El-Kassar, A.-N. (2022) The digital divide: A review and future research agenda. *Technological Forecasting and Social Change* 175, 121359.

Mendez, M.M., Livore, J.P., Calcagno, J.A. and Bigatti, G. (2017) Effects of recreational activities on Patagonian rocky shores. *Marine Environmental Research* 130, 213–220.

Myers West, S. (2018) Censored, suspended, shadowbanned: User interpretations of content moderation on social media platforms. *New Media & Society* 20 (11), 4366–4383.

Napoli, P.M. (2019) Social media and the public interest: Media regulation in the disinformation age. In *Social Media and the Public Interest*. Columbia University Press.

Parcu, P.L., Rossi, M.A., Innocenti, N. and Carrozza, C. (2023) How real will the metaverse be? Exploring the spatial impact of virtual worlds. *European Planning Studies* 31 (7), 1466–1488.

Polishchuk, E., Bujdosó, Z., El Archi, Y., Benbba, B., Zhu, K. and Dávid, L.D. (2023) The theoretical background of virtual reality and its implications for the tourism industry. *Sustainability* 15 (13), Article 13.

Rogerson, C.M. and Saarinen, J. (2018) Tourism for poverty alleviation: Issues and debates in the global south. In The SAGE *Handbook of Tourism Management: Applications of Theories and Concepts to Tourism*. SAGE.

Schutte, N.S. and Stilinović, E.J. (2017) Facilitating empathy through virtual reality. *Motivation and Emotion* 41 (6), 708–712.

Second Life (2024) *Official Site | Second Life—Virtual Worlds,* Virtual Reality, VR, Avatars, and Free 3D Chat. Retrieved from: https://secondlife.com/

Shin, D. (2018) Empathy and embodied experience in virtual environment: To what extent can virtual reality stimulate empathy and embodied experience? *Computers in Human Behavior* 78, 64–73.

Silva, J.N. and Ghilardi-Lopes, N.P. (2012) Indicators of the impacts of tourism on hard-bottom benthic communities of Ilha do Cardoso State Park (Cananéia) and Sonho Beach (Itanhaém), two southern coastal areas of São Paulo State (Brazil). *Ocean & Coastal Management* 58, 1–8.

Steam (2018) *Titanic VR. Steam*. Retrieved from: https://store.steampowered.com/app/741430/Titanic_VR/

tom Dieck, M.C., Jung, T.H. and tom Dieck, D. (2018) Enhancing art gallery visitors' learning experience using wearable augmented reality: Generic learning outcomes perspective. *Current Issues in Tourism* 21 (17), 2014–2034.

Tonazzini, D., Fosse, J. and González, A. (2019) *Blue Tourism—The Transition Towards Sustainable Coastal and Maritime Tourism in World Marine Regions*. Eco-Union . https://doi.org/10.13140/RG.2.2.31175.62885

Trevena, L., Paay, J. and McDonald, R. (2024) VR interventions aimed to induce empathy: A scoping review. *Virtual Reality* 28 (2), 80.

Turvey, S.T. and Crees, J.J. (2019) Extinction in the Anthropocene. *Current Biology* 29 (19), R982–R986.

UN-DESA (2016) *The Sustainable Development Goals Report 2016*. United Nations Department of Economic and Social Affairs.

UNWTO (ed.) (2019) *International Tourism Highlights, 2019 Edition*. World Tourism Organization (UNWTO).

Urry, J. (2015) Climate change and society. In J. Michie and C.L. Cooper (eds) *Why the Social Sciences Matter* (pp. 45–59). Palgrave Macmillan.

van Nuenen, T. and Scarles, C. (2021) Advancements in technology and digital media in tourism. *Tourist Studies* 21 (1), 119–132.

Ventura, S., Badenes-Ribera, L., Herrero, R., Cebolla, A., Galiana, L. and Baños, R. (2020) Virtual reality as a medium to elicit empathy: A meta-analysis. *Cyberpsychology, Behavior, and Social Networking* 23 (10), 667–676.

Verma, S., Warrier, L., Bolia, B. and Mehta, S. (2022) Past, present, and future of virtual tourism-a literature review. *International Journal of Information Management Data Insights* 2 (2), 100085.

Visit Faroe Islands (2017) *SheepView360. Visit Faroe Islands.* Retrieved from: https://visitfaroeislands.com/en/about1/marketing-development-campaigns/sheepview360

Visualise (n.d.) *Thomas Cook Virtual Holiday. Visualise.* Retrieved from https://visualise.com/case-study/thomas-cook-virtual-holiday

Walton, J.K. (2023) *Tourism | Definition, History, Types, Importance, & Industry | Britannica. Britannica.* Retrieved from: https://www.britannica.com/topic/tourism

Wevr (n.d.) *theBlu.* Wevr. Retrieved from: https://wevr.com/theblu

14 Bridging Theory to Practice for an Ethical Animal-Based Tourism Future: The Case of the Italian Association ETICOSCIENZA

Christian Lenzi and Chiara Grasso

Introducing ETICOSCIENZA

In 2018, ETICOSCIENZA Association was founded in Italy. At the conclusion of our university studies in Animal Behavior, we faced the dilemma of either compromising our ethical principles by working in circuses or zoological facilities with limited scientific validity or abandoning our aspiration to work with animals. The scarcity of direct employment links between Italian universities and the workforce, primarily reliant on public tenders or funding, led us to explore alternative paths. In response to this challenge, ETICOSCIENZA was established in May 2018 as a non-profit organization based in Turin. Co-founded by Chiara Grasso (President) and Christian Lenzi (Vice-President), our association aimed to provide students and young naturalists with a contemporary future-oriented perspective and serve as a bridge between academia and employment.

With a background in psychology and wildlife biology we wanted to address the moral issues of ethology from a biocentric perspective, offering an integrated approach between science and the everyday challenges of the relationship between humans and other animals. Our core activities range from training courses to university seminars, from planting new trees to environmental education for children. A central theme that led us to find this association is applied ethology. Pat Monaghan, Regius Professor of Zoology – University of Glasgow, defined it as: '*Behavioural knowledge is essential in dealing with animal problems that result from our exploitation of other animals, competition between ourselves and other species, and in relation to the spread of disease*' (Monaghan, 1984).

ETICOSCIENZA focuses on fostering and advancing the discourse on the moral and ethical implications of applied ethology, drawing inspiration from the foundational work of Marc Bekoff and Jane Goodall and their publication *Ethologists for the Ethical Treatment of Animals* (EETA: Announcement, 2000). Our objective is to mediate between ethics and science, adopting a biocentric inclusive understanding of the human-animal relationship and encouraging critical reflections on animals beyond their 'charisma'. ETICOSCIENZA achieves this through educational activities, scientific dissemination, animal conservation projects, consultation and sustainable animal-based tourism experiences. One of our major subjects of interest is anthrozoology, which studies the relationships between humans and other animals. We are especially interested in the interactions between visitors and facilities housing captive animals, in Italy and the rest of the world. In our projects we continually find ourselves facing the concept of *othering*, with strong discrimination between us and other animals. In addition to people's perception of city wildlife, one thing that strikes us a lot is the attitude of many owners to have wild species at home as pets, forgetting the ethological needs of these animals.

Recognizing the challenges inherent in the interaction and coexistence of humans and non-human animals within tourism contexts – coupled with the increasing interest in animal-based tourism – our association strives to forge a connection between theoretical insights and practical applications, fostering an ethical future for animal-based tourism. Between 2018 and 2023, our association supported various initiatives, including national congresses, university seminars, collaborative scientific research, coordination events for young graduates and website articles (Lenzi *et al.*, 2020; Grasso *et al.*, 2020; Grasso & Lenzi, 2021b). Our daily efforts focus on creating awareness and facilitating an ethical encounter between humans and 'other' non-humans. Central to our activities is scientific communication, exemplified by the participation of Chiara Lenzi in a TEDx talk in 2019 (https://www.ted.com/talks/chiara_grasso_quello_che_non_vi_dicono_sul_turismo_con_gli_animali), shedding light on false sanctuaries in Africa and unethical tourist facilities worldwide.

Our aim is to shift perspectives on animal-based tourism, challenge processes of *othering* in the human-animal relation, and expose the suffering caused by profit-driven practices to envision and build our desired – utopian – animal-based tourism future (Bergman *et al.*, 2010). ETICOSCIENZA has published educational books, such as: 'Viaggia green nella natura' (Grasso & Lenzi, 2021a) –in English *Travel green in nature* and 'I segreti dell'immunità' (Lenzi, 2020) – in English *The secrets of immunity*. In addition, the two essays 'Brutti, sporchi e cattivi ma utilissimi!' (Grasso, 2021) – in English *Ugly, dirty, and mean but very useful*!; 'Fascino bestiale' (Lenzi, 2021) – in English *Animal Charm* in which we elaborate on uncharismatic species and their ecological importance. In each publication, we offer readers a biocentric view, challenging conventional perspectives.

Our headquarter is an educational farm in the Turin hills that serves as a practical avenue for daily education. Here, visitors interact with rescued farm animals and explore meadows, woods and ancient paths. Through these experiences, we aim to impart knowledge and foster awareness, planting seeds of understanding for the protection of our environment.

Anthrozoology and Applied Ethology in Tourism

This section delves into the application of the disciplines of anthrozoology and ethology in animal-based tourism, exploring the concept of charismatic species and its implications for tourism. Anthrozoology is an emerging discipline that studies the relationship between humans and animals, encompassing both domestic and wild species (Siddiq & Habib, 2016). The roots of this interaction trace back to the origin of *Homo sapiens*, who have interacted with numerous animal species from the outset to the present day. As reported by Hosey and Melfi (2014), human-other animals relations go a long way back into prehistory (i.e. more than 50,000 years ago). According to these authors we can distinguish three stages: a pre-domestic era, in which human societies did not perceive a concrete difference between themselves and other animal species; a domestic era, characterized by the development of *othering* differences and human superiority; and a post-domestic era, where most of human societies have direct experience with the animals we eat. From this aspect we can reflect on the different types of human attitudes compared to other species, which range from love (e.g. for companion animals) to persecution (e.g. for pests).

Applied ethology refers to the behavior of non-human animals and to the human activities related to animals, such as pet therapy. With regards to tourism-contexts, the use of anthrozoology and applied ethology is relevant because – among other scopes – it enables a better understanding of tourists' preferences for certain characteristics of animals, whether in the wild or in captivity.

Coming to the central theme of this book, we can say that the entire relationship between humans and other animals is an *othering* process. Our species has always had the innate need, perhaps to better understand its own identity, to differentiate itself from what surrounds it and in this context other species are different from us, companion animals are different from those raised for human consumption and pests living in rural areas are different from charismatic animals with which to take a selfie.

Charisma, which is a pivotal factor in turning an animal into an attraction, is central to recent studies within anthrozoology and applied ethology. The term 'charismatic' initially appeared in scientific literature as a distinctive trait in flagship species identification (Ducarme *et al.*, 2013). Since the early 2000s, 'charisma' has garnered considerable attention in scientific publications on animal conservation (Mazzoldi *et al.*, 2019). This underscores that emotional engagement in the preference for specific animals often underpins also researchers' choices and areas of investigation. Defining the characteristics that render a species charismatic is intricate, but we will enumerate the most salient traits here. Most animal species considered 'charismatic' are large exotic, terrestrial mammals, considered beautiful and impressive. Appearance assumes a critical role; within the framework of the baby-schema concept (Glocker *et al.*, 2009), many species considered charismatic are perceived as endearing. Nevertheless, charisma is not an inherently objective characteristic for animal species since a species may be perceived as charismatic in the Western world – when in captivity – but it may be deemed a pest in its native habitat when roaming freely in the wild (Ducarme *et al.*, 2013). As we observed during our work experience in South Africa, many tourists were attracted by a facility that allowed them to ride elephants. They are animals that are considered charismatic and highly valued by people. But these same animals, free in their natural habitat, are instead persecuted and harassed because they are considered pests, a nuisance to crop producers.

Consequently, the preference for charismatic species often results in a tangible discrimination and processes of *othering* inside and outside tourism. In a recent publication of ours (Tomassini *et al.*, 2022), we articulated concerns about the implications of a distorted perception of the animal world. The human understanding and approach to the animal kingdom pose not only moral dilemmas but also operational concerns, as clearly emerges in tourism-contexts and tourism practices, given the prevalence of charismatic species in zoos, circuses, wildlife parks, wildlife sanctuaries, tourist attractions and safaris (Sherwen & Hemsworth, 2019). Specifically in tourist attractions, what is most striking is this lack of empathy towards other species. Exploring the theme of *othering*, it emerges that in the minds of visitors their own personal needs come first, forgetting the dimension of living beings of the animals used as attractions. Again, humans tend to place a clear and sterile division between themselves and other non-human species.

The Human Fascination for 'Charismatic' Non-Human Animals: Observations from the Field

During the last few decades, the growing interest in wildlife conservation – and its urgency due to severe ongoing biodiversity loss – as well as in human-animal encounters, is evident for all to see. In this context, it is no coincidence that this strong collective emotional drive has led to the establishment of numerous organizations around the world (e.g. World Animal Protection International), also involved in the study of the human perception of animals and the implications in animal-based tourism. This interest in the study of human perception of animals is shared not only by academia and researchers concerned with animal welfare and wildlife conservation, but also by tourism practitioners and facilities involved in forms of animal-based tourism.

Numerous studies have demonstrated that *Homo sapiens* possesses an innate preference for species defined as 'charismatic' for their symbolic value and/or widespread appeal to human beings. Over the years, in fact, researchers have extensively documented how our species is particularly attracted by those sharing a close evolutionary proximity, such as mammals (Smith *et al.*, 2012). However, charisma also serves as a discriminating factor, beyond mere aesthetics and behavior. A charismatic species captures the public attention both in live setting, for instance, in zoological or tourist facilities, and on screen (e.g. in films, cartoons, or advertisements). It is unsurprising that the list of the 20 most charismatic animal species includes iconic animals like large carnivores, koala, giraffes and pandas (Albert *et al.*, 2018). An interesting finding of the researchers is that people's preference for charisma is positively correlated to size: preferred species tend to be larger than others (Berti *et al.*, 2020).

While human preferences towards other species are inherent in our DNA, cultural and geographical factors can significantly influence these preferences. *Othering* attitudes are dynamic and can change over time based on the challenges human societies face. In regions where the conflict between wildlife and humans is most pronounced, such as North America, there is a prevalent negative perception towards large predators (Basak *et al.*, 2023; Kellert *et al.*, 1996). Wolves and bears, for instance, have historically been viewed as adversaries, evoking fear and, ideally, extermination. This perception is not solely based on the actual damage caused by these animals but is also rooted in numerous legends and false beliefs, a phenomenon observed in various species, including bats. However, cultural variations may contribute to differences in the human perception of animal species. For example, peasants or Criollos, and the Indigenous Wichís, residing in the same geographical area (Gran Chaco region of Argentina), exhibit distinct views on wildlife – for example, motivation for hunting as meat acquirement, hunted species and wildlife uses (Camino *et al.*, 2016).

In the collective imagination, the disparity in how one species is perceived compared to another is heavily influenced by their definitions, narratives and historical interactions with humans (Frank *et al.*, 2019). This observation prompts reflection on the consequences of *Homo sapiens*' preference for certain species over others. One such consequence is the impact on participation in conservation programs for endangered species. The presence of charismatic species in a particular geographic area or natural park may attract more funds for environmental protection. Unfortunately, in many cases, the decisive factor is not the actual conservation risk but rather the preference for specific animals. Field studies indicate that the genuine risk of extinction does not consistently motivate donations to conservation programs (Colléony *et al.*, 2017).

Our experiences as researchers and ethologists, particularly in Costa Rica and South Africa, have allowed us to closely examine the main issues of *othering* and human-animal relations. Through two distinct studies (Lenzi *et al.*, 2021; Grasso & Lenzi, 2021a), we found that visitors to wildlife rescue centres or sanctuaries exhibited a strong preference for more charismatic species (e.g. big mammals or cartoon-related iconic animals). Cuteness emerged as the discriminating factor in people's choices, as baby-schema-like animals (e.g. ring-tailed lemur) were frequently selected from respondents as 'cute' species. This preference was also evident in 'hands-off' tourist facilities, where direct contact between humans and wild animals is prohibited. Our field experiences revealed that, while in South Africa, the rule against touching captive animals or disturbing free-roaming animals during safaris is an exception for a few, in Costa Rica, the national government has long banned such negative interactions.

Within the domains of anthrozoology and ethology, examining the ethical interaction between humans and animals reveals noteworthy aspects in the Italian context. Unfortunately, even in our country problems emerge regarding the concept of *othering* (e.g. persecution of local wildlife and detention of exotic species as pets).

In 2024, Italy still permits circus shows featuring wild animals, activities with cetaceans in water parks, visits to zoological facilities where visitors can feed the housed species, and the enrolment in courses to become a falconer. This situation mirrors that of the broader European context, as many laws and regulations are grounded in European Union directives. Another example pertains to the keeping of exotic species as pets, a practice gaining popularity in Italy and unlikely to diminish in the coming years as the desire for snakes, parrots, or meerkats as household companions grows.

The relationship between the Italian human population and wild animals is significantly influenced by millennia of close contact with species indigenous to the geographical area. Wolves, bears, foxes, bats, birds of prey, and snakes have long been subjected to unjustified prejudices rooted in false beliefs and mythological tales. Historical competition for resources, such as hunting or livestock for breeders, has contributed to lingering sentiments of animosity, still evident in some situations today. Unsurprisingly, these animals, along with insects and spiders, are particularly disliked from an early age in Italy (Borgi & Cirulli, 2015).

Conclusions: The Future of Animal-based Tourism via a Utopian Forecast

In conclusion, what are the future scenarios for animal-based tourism? Drawing on our own experience and daily engagement with the activities and initiatives of the association ETICOSCIENZA, we can identify two emerging trends that most likely will shape the future of animal-based tourism. On the one hand, we observe a shift in the expectations and demands of tourists with regards to the encounter and interaction with animals. More specifically, there is a growing attention to respecting animals and nature, with a growing number of tourists and daily visitors making conscious choices during their travels or nature excursions. Such growing awareness for an ethical animal-based tourism will be strengthened in the forthcoming years, by novel directives and initiatives at European level the European Union focusing on animal welfare and wildlife conservation fostering the request for an ethical animal-based tourism.

On the other side, we also observe a persistent trend and preference for charismatic animal species. This inclination has been embedded in our human species since its origin and is likely to continue shaping our perception of the animal world in the future, influencing our interest and commitment to their protection. Consequently, we envision the future of animal-based tourism with a specific hope – that the enduring human preference for charismatic species will translate into funding for entire ecosystems, encompassing the less charismatic species as well. Simultaneously, we anticipate that all tourist facilities hosting rescued animals will include endangered species indiscriminately, transcending considerations of charisma and aesthetics. This approach aims to raise public awareness and mitigate the risk of extinction for fauna most vulnerable to human impact. In envisaging these new futures, it becomes crucial to prioritize comprehensive ecosystem conservation and foster a paradigm shift towards appreciating the intrinsic value of all species, beyond their charisma and appeal to human beings.

Drawing on the utopian forecast proposed by Bergman *et al.* (2010), we can imagine our utopian future, grounded in the concept of *othering* discussed in this chapter. In doing so, we envision a future where a biocentric view of human-animal relations becomes more widespread. This shift would manifest not only in tourism but also in educational systems and broader societal interactions with nature. In environmental education, working with children to shape their views of the natural world could help reduce species discrimination, fostering a society where coexistence with biodiversity is the norm. While this future is optimistic, it will require concerted efforts to bridge the gap between human preferences and the urgent need for holistic conservation. If we successfully integrate the ethical, ecological and educational dimensions of this vision, we can hope to achieve a utopia where all species are valued and protected.

References

Albert, C., Luque, G.M. and Courchamp, F. (2018) The twenty most charismatic species. *PloS One* 13 (7), e0199149.

Announcement (2000) *Journal of Applied Animal Welfare Science* 3 (3), 277. http://doi.org/10.1207/S15327604JAWS0303_11

Basak, S.M., Rostovskaya, E., Birks, J. and Wierzbowska, I.A. (2023) Perceptions and attitudes to understand human-wildlife conflict in an urban landscape–A systematic review. *Ecological Indicators* 151, 110319.

Bergman, A., Karlsson, J.C. and Axelsson, J. (2010) Truths claims and explanatory claims – an ontological typology of futures studies. *Futures* 42 (8), 857–865.

Berti, E., Monsarrat, S., Munk, M., Jarvie, S. and Svenning, J.C. (2020) Body size is a good proxy for vertebrate charisma. *Biological Conservation* 251, 108790.

Borgi, M. and Cirulli, F. (2015) Attitudes toward animals among kindergarten children: Species preferences. *Anthrozoös* 28 (1), 45–59.

Camino, M., Cortez, S., Cerezo, A. and Mariana, A. (2016) Wildlife conservation, perceptions of different co-existing cultures. *International Journal of Conservation Science* 7 (1), 109–122.

Colléony, A., Clayton, S., Couvet, D., Saint Jalme, M. and Prévot, A.C. (2017) Human preferences for species conservation: Animal charisma trumps endangered status. *Biological Conservation* 206, 263–269.

Ducarme, F., Luque, G.M. and Courchamp, F. (2013) What are "charismatic species" for conservation biologists. *BioSciences Master Reviews* 10 (2013), 1–8.

ETICOSCIENZA. https://www.eticoscienza.it (accessed 14 July 2023).

Frank, B., Glikman, J.A. and Marchini, S. (eds) (2019) *Human–Wildlife Interactions: Turning Conflict into Coexistence* (Vol. 23). Cambridge University Press.

Glocker, M.L., Langleben, D.D., Ruparel, K., Loughead, J.W., Gur, R.C. and Sachser, N. (2009) Baby schema in infant faces induces cuteness perception and motivation for caretaking in adults. *Ethology* 115 (3), 257–263.

Grasso, C., Lenzi, C., Speiran, S. and Pirrone, F. (2020) Anthropomorphized nonhuman animals in mass media and their influence on human attitudes toward wildlife. *Society & Animals* 1 (aop), 1–25.

Grasso, C. (2021) *Brutti, Sporchi e Cattivi ma Utilissimi!* Edizioni Lindau.

Grasso, C. and Lenzi, C. (2021a) *Viaggia Green Nella Natura*. Edizioni Sonda.

Grasso, C. and Lenzi, C. (2021b) Preliminary study of tourists' preferences in a Costa Rican wildlife sanctuary. https://www.researchgate.net/publication/353555414_Preliminary_study_of_tourists'_preferences_in_a_Costa_Rican_wildlife_sanctuary

Hosey, G. and Melfi, V. (2014) Human-animal interactions, relationships and bonds: A review and analysis of the literature. *International Journal of Comparative Psychology* 27 (1), 117–142.

Kellert, S.R., Black, M., Rush, C.R. and Bath, A.J. (1996) Human culture and large carnivore conservation in North America. *Conservation Biology* 10 (4), 977–990.

Lenzi, C., Speiran, S. and Grasso, C. (2020) "Let me take a selfie": Implications of social media for public perceptions of wild animals. *Society & Animals* 31 (1), 64–83.

Lenzi, C. (2020) *I Segreti Dell'immunità*. Edizioni Lindau.

Lenzi, C. (2021) *Fascino Bestiale*. Edizioni Lindau.

Lenzi, C., Grasso, C. and Speiran, S. (2021) Visitors' attitudes toward non-human primates in a free-roaming multi-species sanctuary (Monkeyland, South Africa). https://www.preprints.org/manuscript/202107.0684/v1

Mazzoldi, C., Bearzi, G., Brito, C., Carvalho, I., Desiderà, E., Endrizzi, L. ... and MacDiarmid, A. (2019) From sea monsters to charismatic megafauna: Changes in perception and use of large marine animals. *PLoS One* 14 (12), e0226810.

Monaghan, P. (1984) Applied ethology. *Animal Behaviour* 32 (3), 908–915.

Sherwen, S.L. and Hemsworth, P.H. (2019) The visitor effect on zoo animals: Implications and opportunities for zoo animal welfare. *Animals* 9 (6), 366.
Siddiq, A.B. and Habib, A. (2016) Anthrozoology–An emerging robust multidisciplinary subfield of anthropological science. *Green University Review of Social Sciences* 3 (1), 45–67.
Smith, R.J., Veríssimo, D., Isaac, N.J. and Jones, K.E. (2012) Identifying Cinderella species: Uncovering mammals with conservation flagship appeal. *Conservation Letters* 5 (3), 205–212.
TEDx talk of Chiara Grasso. https://www.ted.com/talks/chiara_grasso_quello_che_non_vi_dicono_sul_turismo_con_gli_animali (accessed 14 July 2023).
Tomassini, L., Bertella, G., Grasso, C. and Lenzi, C. (2022) The space of animal justice in wildlife sanctuaries: A posthuman perspective. *Journal of Ecotourism* 1–18. 23 (1), 38–55.
World Animal Protection website. https://www.worldanimalprotection.org/ (accessed 14 July 2023).

15 Conclusive Reflections on Tourism Futures: Memos, Turning Points and Emerging Themes

Lucia Tomassini

This book has presented a critical reflection on the space of human and non-human actors in tourism futures by focusing on the notion of otherness and processes of *othering*. The exploration of theoretical perspectives on space, otherness, and processes of *othering* forms the core of understanding tourism's evolving qualities and values in this new millennium. Each author has engaged with this overarching theme by examining different aspects and contexts of the spaces occupied by human and non-human actors in tourism futures, discussing the implications for tourism's future and envisioning potential scenarios.

The point of departure of this book (Chapter 1) drew on Calvino's (1988) *Six Memos for the Next Millennium*, posthuman theory (Braidotti, 2011, 2013), and Massey's (2005) ruminations on space to explore the intersection of tourism and its futures. It acknowledges the spatial entanglement of human and non-human actors within tourism contexts, emphasizing the need to envision an open and relational understanding of space in tourism's future. More specifically, Chapter 2 examined the values and qualities of tourism as it transitions into this new millennium, drawing inspiration from Calvino's (1988) memos of *lightness, quickness, exactitude, visibility* and *multiplicity*. The chapter presents and discusses *lightness* as 'relational sustainability' and 'metamorphosis', *quickness* as 'Festina Lente', *exactitude* as the capacity to embrace and convey complexity with clarity and accuracy, *visibility* as 'imagineering' and *multiplicity* as an 'open eventful and relational spatiality'. Hence, drawing on the spatial turn in the social sciences (Massey, 2005) and posthuman affirmative ethics (Braidotti, 2011, 2013), it elaborates on the qualities and key themes of contemporary tourism concerning otherness and processes of *othering* as it transitions into this new millennium.

This perspective runs throughout all the chapters of the book, which collectively invite readers to unpack contemporary tourism by recognizing the challenges of a tourism space imbued with multiple entanglements among human and non-human others, as well as characterized by uneven processes of *othering*. As such, this conclusive chapter examines how each book chapter contributes to and enrich Calvino's future-oriented values (1908) – *lightness, quickness, exactitude, visibility, multiplicity* – helping to identify, describe and examine key themes of the contemporary tourism and the implications for its future development (Table 15.1).

This edited book drew on future studies and theories concerning the future turning points of tourism development (Yeoman & McMahon-Beattie, 2020), typologies of futures studies (Bergman *et al.*, 2010) and critical futures research approaches (Inayatullah, 1998). By acknowledging that the future of tourism will unfold in a volatile, uncertain, complex, and ambiguous world, this book emphasizes the importance of tourism futures research, scenario awareness, and the ability to foresee possible directions for tourism's evolution (Yeoman & McMahon-Beattie, 2023). Table 15.2 offers an overview of the interplay between the book's chapters and the three future turning points of tourism development (Yeoman & McMahon-Beattie, 2020).

A key insight from Table 15.2 is that **Future Turning Point 1: The Political and Economic Importance of Tourism** emerges as the dominant theme across most chapters. This is unsurprising given that otherness, othering, and spatial justice are deeply tied to governance, economic structures and the sociopolitical dimensions of tourism. Many chapters critically explore issues such as migration and more-than-human and multispecies justice, reinforcing the centrality of political and economic discourses in shaping tourism futures. **Future Turning Point 2: Tourism Footprint** is less prominent but remains significant, particularly in discussions on sustainability, regenerative tourism and ethical interactions with non-human actors. Chapters on marine conservation, rural tourism and digital tourism address the environmental and – to some extent – social impacts of tourism activities, aligning with sustainability concerns. Finally, **Future Turning Point 3: Transformational Technologies** is strongly represented in fewer chapters but plays a crucial role where discussed. Chapters focusing on immersive technologies, digital worlds, and digital storytelling demonstrate how technological advancements are reshaping tourism experiences and governance. As such, technology remains a mediating rather than a dominant theme, often intersecting with the other two turning points.

In conclusion, the future turning points emerging in this book – namely political-economic shifts, tourism's evolving environmental footprint, and the emergence of transformational technologies – reveal tourism as an inherently entangled and contested space of future-making. Hence, the chapters collectively invite a critical rethinking of how we

Table 15.1 Chapters' contribution and interplay with Calvino's memos (1988)

Chapter	*Title*	*Lightness*	*Quickness*	*Exactitude*	*Visibility*	*Multiplicity*
	PART I: Theoretical perspectives on otherness, processes of othering, and space in contemporary tourism and implications for its future development					
3	A posthuman theorisation of otherness and othering in tourism and migration (Burrai)	*Being human as a posthuman process of becoming*			*Hybrid communities with a posthuman imaginative culture*	*Intersectionality and affirmative ethics for relational open tourism futures*
4	Animal ethics and cultural relativism: justice for non-human animals in tourism futures (Rizzolo & Fennell)			*Images and languages for ethical tourism framework and justice-oriented policies*	*Representing and valuing non-human agency*	Multispecies justice and intersectional more-then-human ethics
5	Sugar shaking tourism assemblage: deterritorialization and *othering* maples trees in complex tourism futures (Lapointe *et al.*)	*Assemblage theory for a tourism future with 'light' impacts on more-than-human host communities*		*Destination management and governance for territorial identity and cultural sustainability*	*Challenging heritage commodification and reterritorialization*	
	PART II: Navigating spatial entanglements with the 'other' and implications for tourism futures					
6	The future of rural tourism and leisure literature: perpetuating or challenging the 'othering' of farm animals? (Bertella & Lombardini)			*Developing a coherent and structured narrative and research approach to farm animals in tourism*	*Constructing a novel gaze for farm animals and farm tourism challenging uneven power relations*	*Multispecies interactions and more-than-human affirmative socio-spatial entanglements*
7	Human and non-human animal mobilities in an insecure world: a critique of (de)humanizing narratives and implications for alternative futures (Muldoon)			*Challenging the current mobility discourse and narrative for an inclusive understanding of the 'other'*		*An open relational spatiality and mobility for an open inclusive tourism space*
8	Encountering the other: the role of underwater photographers in shaping the futures of regenerative marine tourism (Hoarau-Heemstra & Wally Ryan)	*Regenerative marine tourism for transformative learning and emotional attachment*		*Precision and accuracy in marine conservation narratives and management practices*	*Visual storytelling for regenerative marine tourism*	*Spatial and temporal multispecies connections: a posthuman approach to marine tourism*

Table 15.1 (*Continued*)

Chapter	*Title*	*Lightness*	*Quickness*	*Exactitude*	*Visibility*	*Multiplicity*
9	Video methods in co-creating space for non-human animals in tourism futures (Äijällä *et al.*)		*Technological mediation for co-created digital storytelling*		*Co-creation of narratives through animal agency for ethical representation*	*Participatory novel methodologies for constructive multispecies entanglements*
10	Exploring the Twilight Zone: Anthropology, Science Fiction and Tourism Futures in Urban Environments (Vietti)	*Transformative solarpunk tourism in urban space*	*Solarpunk tourism through technological integration*		*Sci-fi imaginaries for alternative solarpunk future of tourism*	*Multidisciplinary relational approaches and methodologies for a more-than-human tourism future*
				PART III: Views and insights from independent researchers and practitioners		
11	Tourism, urban population, and *othering*: critical reflections and forecasts for Italian cultural cities (Gainsforth)	*Reducing and managing tourism impacts on the physical and sociological space of cultural cities*		*Policies and governance to manage and plan urban population and tourism in cultural cities*	*Imaginative culture in the evolving scenario of cultural cities*	
12	The space of young residents with a migrant background as cross-borders guides: the project 'Rural Migrantour: Paths of Recovery' (Milovanovič)	*Transformative and inclusive cultural tourism in cross-border territories*			*Migration narratives and gazes, and cross-cultural mediation in cross-border territories tourism*	*Transcultural identities and inclusive hybrid identities*
13	Digital Worlds for Blue Tourism: the case of extended realities in tourism futures (Lovecky & Mueller)	*Virtual tourism and immersive technologies for lower tourism impact and transformative learning*	*Novel, immersive technologies in blue tourism futures*		*Digital worlds and novel landscapes for marine environment conservation*	
14	Bridging theory to practice for an ethical animal-based tourism future: the case of the Italian association ETICOSCIENZA (Lenzi & Grasso)			*Animal welfare policies and ethical tourism frameworks*	*Inclusive and equitable non-human representation in tourism*	*More-than-human biocentric tourism models and approaches for an interdisciplinary ethics*

Table 15.2 Chapters' relation with Future Turning Points in tourism development (Yeoman & McMahon-Beattie, 2020)

Chapter	Title	Future Turning Point 1: *Political & Economic importance of Tourism*	Future Turning Point 2: *Tourism Footprint*	Future Turning Point 3: *Transformational Technologies*
3	A posthuman theorisation of otherness and othering in tourism and migration (Burrai)	✓		
4	Animal ethics and cultural relativism: justice for non-human animals in tourism futures (Rizzolo & Fennell)	✓	✓	
5	Sugar shaking tourism assemblage: deterritorialization and *othering* maples trees in complex tourism futures (Lapointe *et al.*)	✓	✓	
6	The future of rural tourism and leisure literature: perpetuating or challenging the 'othering' of farm animals? (Bertella & Lombardini)	✓		
7	Human and non-human animal mobilities in an insecure world: a critique of (de)humanizing narratives and implications for alternative futures (Muldoon)	✓		
8	Encountering the other: the role of underwater photographers in shaping the futures of regenerative marine tourism (Hoarau-Heemstra & Wally Ryan)	✓	✓	✓
9	Video methods in co-creating space for non-human animals in tourism futures (Äijällä *et al.*)	✓	✓	✓
10	Exploring the Twilight Zone: Anthropology, Science Fiction and Tourism Futures in Urban Environments (Vietti)	✓	✓	✓
11	Tourism, urban population, and *othering*: critical reflections and forecasts for Italian cultural cities (Gainsforth)	✓		
12	The space of young residents with a migrant background as cross-borders guides: the project 'Rural Migrantour: Paths of Recovery' (Milovanovič)	✓		
13	Digital Worlds for Blue Tourism: the case of extended realities in tourism futures (Lovecky & Mueller)	✓	✓	✓
14	Bridging theory to practice for an ethical animal-based tourism future: the case of the Italian association ETICOSCIENZA (Lenzi & Grasso)	✓	✓	

theorize and intervene in the becoming of tourism. The book foregrounds an ethics of relationality that cuts across more-than-human justice, multispecies encounters, digital mediation, and spatial complexity. Through this lens, tourism futures are not only about reacting to trends or forecasting markets but about interrogating foundational assumptions regarding who and what constitutes tourism and its stakeholders. By integrating posthuman theory and spatial reflections, the chapters demonstrate how tourism is continuously shaped by intersecting material, political, and affective entanglements challenging the notion of otherness and processes of *othering*. For instance, discussions on multispecies justice, marine regenerative practices and solarpunk imaginaries go beyond critique to offer affirmative visions of coexistence. These are not merely conceptual add-ons but theoretical reflections that demand a profound epistemological and ontological expansion of tourism studies. The multiplicity of perspectives across the book reflects a field in transformation, one that must critically account for both the planetary urgencies and the micro-scale ethics of tourism encounters, which are often more-than-human.

This edited book also contributes to the theoretical scaffolding of tourism futures by connecting emerging key themes and practices to broader paradigms such as posthuman nomadic subjectivity (Braidotti, 2011, 2013), multiple open relational spatiality (Massey, 2005), and critical futures methodology (Inayatullah, 1998). These connections allow for a richer understanding of how future tourism might navigate a volatile, uncertain, complex, and ambiguous world without defaulting to managerial or technocratic solutions. The chapters reveal how imaginaries – whether through digital storytelling, speculative fiction, or visual media –play a crucial role in shaping what is thinkable, desirable, and possible within tourism. Moreover, the emphasis on turning points is not simply to chronicle change but to highlight the moments where ethical, political and technological ruptures generate space for alternative futures. In doing so, the volume offers a layered and reflexive contribution to futures studies in tourism – situating itself not just as a thematic collection of contributions but as a manifesto for critical engagement. Future scholarship and policymaking can build on these insights, developing policies and practices that are inclusive, post-anthropocentric and deeply attentive to the interplay between more-than-human justice, space and technology. In sum, this conclusion affirms that the future of tourism depends not only on understanding emerging trends, but on cultivating the theoretical and ethical agility to imagine and co-create futures that are equitable, plural, inclusive and socially and environmentally responsive.

References

Bergman, A., Karlsson, J.C. and Axelsson, J. (2010) Truths claims and explanatory claims – An ontological typology of futures studies. *Futures* 42 (8), 857–865.

Braidotti, R. (2011) *Nomadic Theory: The Portable Rosi Braidotti*. Columbia University Press.
Braidotti, R. (2013) *The Posthuman*. John Wiley & Sons.
Calvino, I. (1988) *Six Memos for the Next Millennium*. Harvard University Press.
Massey, D. (2005) *For Space*. Sage.
Inayatullah, S. (1998) Causal layered analysis: Poststructuralism as method. *Futures* 30 (8), 815–829.
Yeoman, I. and McMahon-Beattie, U. (2020) Turning points' in tourism's development: 1946–2095, a perspective article. *Tourism Review* 75 (1), 86–90.
Yeoman, I. and McMahon-Beattie, U. (2023) Future past of tourism: Critical reflection's on the rise of tourism futures. *Tourism Geographies* 27 (3&4), 476–492. https://doi.org/10.1080/14616688.2023.2270937

Index

www.ingramcontent.com/pod-product-compliance
Lightning Source LLC
Chambersburg PA
CBHW072040051025
33612CB00003B/26
* 9 7 8 1 8 4 5 4 1 9 3 2 5 *